Moving Things

Moving Things

WALLSTEIN VERLAG

Inhalt

Peter J. Bräunlein   7
Moving Things – Einleitung

Maliheh Bayat Tork | Antonie Fuhse | Andrea Lauser |
Friedemann Yi-Neumann | Peter J. Bräunlein | Joachim Baur   21
»Moving things« erforschen: Ein Gespräch
des Projektteams über Herausforderungen und Chancen

Friedemann Yi-Neumann   37
Foodways

Elza Czarnowski   63
Schlüssel

Özlem Savaş   75
Bücher auf dem Weg nach Hause

Veronika Reidinger | Anne Unterwurzacher   93
Vom Koffer zum Fluchtgepäck und wieder zurück

Romm Lewkowicz   115
Die Papierspur der Migration:
Fragmente einer Ethnographie des Reisepasses

Peter J. Bräunlein   135
Von Puppen, Bären, Drachen und Waffen:
Spielzeug, Flucht und Migration

Anoush Masoudi   159
Schuhe: Drei Schritte eines langen Weges

Maliheh Bayat Tork | Antonie Fuhse   181
Menstruationsprodukte

Andrea Lauser | Miriam Kuhnke |
Maliheh Bayat Tork | Antonie Fuhse   201
Das Smartphone: Überlebenswerkzeug und mobile Heimat

Nina de la Chevallerie   231
Die Objektivierung von Migration
in den Darstellenden Künsten: Eine Annäherung des freien
Theaters »boat people projekt« aus Göttingen

Autor:innen   260

Impressum   264

Content

7	Peter J. Bräunlein Moving Things – An Introduction
21	Maliheh Bayat Tork | Antonie Fuhse | Andrea Lauser | Friedemann Yi-Neumann | Peter J. Bräunlein | Joachim Baur Researching "Moving Things": A Project Team Discussion About Challenges and Chances
37	Friedemann Yi-Neumann Foodways
63	Elza Czarnowski Keys
75	Özlem Savaş Books Heading Home
93	Veronika Reidinger | Anne Unterwurzacher From Suitcase to Flight Baggage and Back
115	Romm Lewkowicz The Paper Trail of Migration: Fragments of an Ethnography of the Passport
135	Peter J. Bräunlein On Dolls, Bears, Kites and Weapons: Toys, Flight and Migration
159	Anoush Masoudi Shoes: Three Steps in a Long Journey
181	Maliheh Bayat Tork | Antonie Fuhse Menstruation Products
201	Andrea Lauser | Miriam Kuhnke | Maliheh Bayat Tork | Antonie Fuhse The Smartphone: Survival Tool and Mobile Sense of Home
231	Nina de la Chevallerie The Objectivisation of Migration in the Performing Arts: A Joint Publication from the Independent Theatre Company "boat people projekt" from Göttingen, Germany
260	Authors
264	Imprint

Moving Things – Einleitung
Moving Things – An Introduction

Peter J. Bräunlein

Abb. 1: Javads Schnürsenkel, 2020. ©Javad

Image 1: Javad's shoelaces, 2020. ©Javad

Diese Schuhe gehören Javad. Die weißen Schnürsenkel in den Sneakers sind nicht die Originale. Die ursprünglichen Schnürsenkel waren grau, die gleiche Farbe wie die Schuhe. Sie wurden von der Grenzpolizei in Griechenland konfisziert. In einem Chatgespräch mit Javad erzählte er mir von den fünf Tagen, die er, nachdem er vom Iran zunächst in die Türkei gelangt war, in seinem eigentlichen Zielland Griechenland ver-

These are Javad's shoes. The white shoelaces in the trainers are not those that originally came with the footwear. They were grey, just like the shoes. And they were confiscated by the border police in Greece. In an internet chat, Javad told me about the five days he spent in Greece – his original destination – after he had already journeyed from Iran to Turkey. "The

brachte. »Die fünf Tage in Griechenland waren sehr hart. Wir liefen in der Nacht und schliefen am Tag. Uns gingen das Essen und das Wasser aus, aber wir liefen weiter. Fünf Tage lang Elend und am Ende abgeschoben. Die Polizei nahm uns alles, unsere Handys, Rucksäcke und Kleidung.« »Nahmen sie euch auch euer Geld?«, fragte ich. »Nein, sie gaben uns unser Geld zurück und auch einen Satz Kleidung. Den Rest nahmen sie mit.« Ich fragte, warum die Polizei ihre Handys konfisziert hatte. »Um zu verhindern, dass wir es noch einmal versuchen. Alle benutzen das GPS auf ihren Handys, um die Routen zu finden, und die Polizei weiß das. Sie nahmen auch unsere Schnürsenkel und Gürtel, um uns zu zwingen aufzugeben. Ohne diese Dinge kann man nicht weit laufen.« Ich hatte nie realisiert, welche große Rolle diese kleinen Dinge, wie Schnürsenkel oder Gürtel, auf der Flucht spielen und wie der Verlust dieser Dinge die Mobilität der Menschen ernsthaft einschränkt. Deswegen fragte ich Javad danach, wie er sich fühlte, als er das Handy, die Schnürsenkel und den Gürtel an den Polizisten übergeben hatte. Er antwortete: »Ich trauerte um mein Handy, es war ein Geschenk von meinem Bruder. Ich liebte es. Aber die anderen Dinge waren egal.« Er erzählte weiter: »Meine Schnürsenkel waren grau, sie hatten die gleiche Farbe wie die Schuhe. Als ich wieder in Istanbul war, suchte ich nach grauen Schnürsenkeln, konnte aber keine finden. So kaufte ich weiße. Sie passen nicht zur Farbe meiner Schuhe, und so erinnere ich mich jedes Mal beim Binden und Öffnen meiner Schuhe an die Tage in Griechenland.« Ich wollte wissen, welche Gefühle er damit verband: Wut, Scham, Reue, Demütigung, Traurigkeit ...? »Nichts davon! ... Es ist tiefes Bedauern. Wir hatten die Grenze überschritten, wir waren in Griechenland, wir waren unserem Ziel so nah! Ich wünschte, wir wären

five days in Greece were extremely tough. We walked by night and slept by day. We ran out of food and water, but we kept on walking. Five days of misery, and then we were deported at the end of it. The police took everything from us, our mobile phones, backpacks and clothing." "Did they also take your money?" I asked. "No, they gave us our money back along with one set of clothes. They took everything else away with them." I asked why the police had confiscated their mobile phones. "To prevent us attempting it a second time. The police knew that everyone was using GPS on their phones to find the routes. They also took away our shoelaces and our belts, to force us to give up. You can't walk far without these things." Up to this point, I had not realised what a major role small things like shoelaces or belts play for people who are fleeing, and how losing these things can substantially reduce their mobility. This prompted my next question about how he had felt after he had handed over his phone, shoelaces and belt to the police officers. He replied, "I grieved for my mobile phone because it was a present from my brother. I loved it. But I didn't care about the other things." He continued, "My shoelaces were grey, the same colour as my shoes. When I got back to Istanbul, I looked for grey shoelaces, but couldn't find any. So I bought white ones. They don't match the colour of my shoes, and so every time I tie and untie my laces, I'm reminded of my days in Greece." I wanted to know which feelings he connected with this experience: anger, shame, remorse, humiliation, or sadness ...? Javad's response was:

nicht verhaftet worden und hätten es nach all der Mühe geschafft«, antwortete Javad.

Diese Geschichte dokumentierte Maliheh Bayat Tork, Mitarbeiterin in unserem Forschungsprojekt »Zur Materialität von Flucht und Migration«.[1] Maliheh lernte Javad in einem Flüchtlingslager im Iran kennen. Als Sohn afghanischer Eltern im Iran aufgewachsen, hatte er eine Beschäftigung im Tätigkeitsfeld IT gefunden. Als der Iran in den Jahren 2018/19 aufgrund von internationalen Sanktionen eine Wirtschaftskrise durchlebte, wurde sein Gehalt gestrichen. Nach einjähriger Arbeitslosigkeit und ohne Aussicht auf eine bezahlte Tätigkeit, entschloss er sich zu Beginn des Jahres 2020, das Land zu verlassen, und machte sich auf den Weg nach Europa. Weite Teile der Strecke legte er dabei zu Fuß zurück.

Das Foto von Javads Schuhen lenkt unseren Blick auf eine elementare Dimension von Migration, der bislang zu wenig Aufmerksamkeit geschenkt wurde, nämlich auf das Verhältnis von materieller Kultur und Migration. Schnürsenkel scheinen uns zunächst als triviale Objekte. Zwar sind auch in unserem Alltag Schnürsenkel durchaus wichtig und ihr Fehlen wäre hinderlich. Doch wir schenken dieser Tatsache keine Aufmerksamkeit. Für eine Person, die sich auf der Flucht befindet, sieht dies anders aus. Javads Blick auf seine Schnürsenkel regt uns an, seinem Blick zu folgen und Dinge wichtig zu nehmen.

Zudem, so wird aus dem Bericht Javads deutlich, speichern Dinge Erinnerung und sind gefühlsgeladen. So beschreibt er seine Schnürsenkel (und auch sein entwendetes Smartphone) nicht nur funktional, sondern

"None of them! ... It is more like deep regret. We had crossed the border, we were so near to our goal. I wished that we would not be arrested, and that we would make it after all our efforts."

This story was documented by Maliheh Bayat Tork, a staff member in our research project "On the Materiality of (Forced) Migration".[1] Maliheh got to know Javad in a refugee camp in Iran. Javad, who grew up as the son of Afghani parents in Iran, had worked in the IT sector before the economic crisis caused by international sanctions hit the country in 2018 and 2019. Consequently, his salary was cut. After a year of unemployment and with no prospects of a new paid job, in early 2020, he decided to leave Iran and make his way to Europe. He covered large stretches of his migration route on foot.

The photo of Javad's shoes pulls our gaze towards one elementary dimension that has not yet received the attention it deserves – the relationship between material culture and migration. Shoelaces at first appear to us to be trivial objects; while they are certainly important in our everyday lives, and we would be inconvenienced by their absence, we do not initially attach significance to this fact. But for a person who is fleeing, things look very different. The way Javad sees his shoelaces can prompt us to follow his gaze and take things seriously – entities we have not sufficiently considered until now.

Moreover, Javad's report makes evident that things act as reservoirs of memory and are laden with feelings. Thus, he describes his shoelaces and his confiscated

Moving Things – An Introduction

vor allem emotional (siehe hierzu auch den Beitrag zum Smartphone von Andrea Lauser, Miriam Kuhnke, Maliheh Bayat Tork und Antonie Fuhse in diesem Band). Javads Schnürsenkel erzählen somit von der Mühsal der Migration, von der Unbarmherzigkeit des europäischen Grenzregimes, von vergeblicher Hoffnung und der Erfahrung des Scheiterns.

Kurz: Im Kontext von Flucht und Migration kommt dem Mensch-Ding-Verhältnis besondere Bedeutung zu. Darum geht es uns im vorliegenden Band, der im Rahmen des mehrjährigen Forschungsprojektes »Zur Materialität von Flucht und Migration« entstanden ist.[2] Mit unserem Blick auf bewegte und bewegende Dinge möchten wir Impulse für die Migrationsforschung liefern und darüber hinaus einen Zugang zu einem erweiterten Verständnis von Flucht und Migration eröffnen.

In der Gesellschaft von Dingen leben

Wir alle pflegen intensive Beziehungen zu Dingen, wir leben in der Gesellschaft von Dingen. Dinge von Verstorbenen werden sorgsam aufbewahrt, Computer nicht selten beschimpft, einzelne Kleidungsstücke, Kunstobjekte oder Bücher innig geliebt. Manche geben ihrem Motorrad einen Kosenamen, andere sprechen mit ihrem Auto, viele streicheln ihr Smartphone. Manche Dinge sind uns buchstäblich sehr nahe. Der Fingerring etwa, die Halskette oder der Piercingschmuck. Andere Dinge sind gar Teil unseres Körpers geworden: der Herzschrittmacher, die Hüftprothese oder das Brustimplantat.

smartphone not in a merely functional way but primarily in an emotional manner (see also the essay by Andrea Lauser, Miriam Kuhnke, Maliheh Bayat Tork and Antonie Fuhse in this volume). This enables Javad's shoelaces to narrate the tribulations of migration, the hard-heartedness of Europe's border regime, the act of hoping – sometimes unavailingly – and the experience of failure.

To sum up: in the context of flight and migration, relations between people and things acquire special significance. This is the subject of this volume, which has taken shape within the framework of the multiyear research project "On the Materiality of (Forced) Migration".[2] With our research focus on moved and moving things, we wish to offer new impetus for migration research. Beyond this, we wish to expand access to an enriched understanding of flight and migration.

Life in the Society of Things

We all live in the society of things, we all nurture intensive relationships with these entities. Things belonging to the dead are carefully kept, computers get sworn at on a regular basis, and individual pieces of clothing, works of art or books are loved rapturously. Some people give their motorbike a nickname, others have conversations with their cars, and many individuals can be seen stroking their smartphones. Some things are literally very close to us. A ring on our finger, for example, or a necklace, or a piercing jewellery. Other things even be-

Viele Dinge dienen einem praktischen Zweck, doch manchmal bleibt ihre Funktion scheinbar im Dunklen. So begehren und horten Sammler weitgehend nutzlose Dinge, doch auch in Durchschnittshaushalten westlicher Gesellschaften häufen sich massenhaft Dinge, die nicht wirklich gebraucht werden. Das Praktische und Instrumentelle bestimmt nur einen Teil der Mensch-Ding-Beziehung. Über Dinge werden Individualität und gesellschaftlicher Status signalisiert, sie ermöglichen den Anschluss an eine Gemeinschaft und erleichtern die Kommunikation über die Distanz hinweg. Eigene Wünsche und Vorstellungen sind ebenso im Spiel wie Emotionen.

Dinge greifen mehr in unser Leben ein, als wir vermuten.

Der britische Ethnologe Daniel Miller stellt in seinen Forschungen zu Kleidung, Wohnen, Smartphone- und Internetgebrauch die These auf, dass Menschen erst durch die Aneignung von Dingen zu kulturellen Subjekten werden. Aktiver Umgang mit der Dingwelt sei der Weg, auf dem Menschen Kultur, also soziale Strukturen, Ideen, Normen, Werte und Handlungsmuster, internalisieren und inkorporieren. Menschen würden etwas mit Dingen tun, umgekehrt würden Dinge aber auch etwas mit den Menschen machen (Miller 2008: 272).

> Wir können nicht wissen, wer wir sind oder werden, was wir sind, außer durch den Blick in einen materiellen Spiegel, der die historische Welt ist, die von denen geschaffen wurde, die vor uns lebten. Diese Welt tritt uns als materielle Kultur entgegen und entwickelt sich durch uns weiter. (Miller 2005: 8, Übersetzung PJBr)

come a part of our own body: a pacemaker, an artificial hip or breast implants.

While many things serve a practical purpose, at times their functions seem to remain in the dark. Collectors desire and hoard largely useless things, but average households in Western societies contain huge piles of stuff that is not really needed. Practical and instrumental facets define only one part of human-thing relations. Things are also used to signalise individuality and societal status, making participation in particular communities possible and easing communication over large distances. The possessors' wishes and notions have their role to play in what things are, just as emotions also have theirs.

Things intervene in our lives more than we might suspect.

The British social anthropologist Daniel Miller presents the thesis, as an adjunct to his research on clothing, housing and smartphone and internet use, that people only become cultural subjects through the appropriation of things. In this view, humans internalise and incorporate culture – social structures, ideas, norms, values and behaviour patterns – through active engagement with the world of things. People carry out actions with and to things, but the reverse is also true: Things also do something to people (Miller 2008: 272).

> We cannot know who we are, or become what we are, except by looking in a material mirror, which is the historical world created by those who lived before us. This world confronts us as material culture and

Den Gedanken, dass Dinge etwas mit Menschen machen, spitzt der französische Soziologe und Philosoph Bruno Latour zu. Dass der Mensch Herr über die Dinge ist, die er doch selbst geschaffen hat, sei eine »hübsche Geschichte, doch sie kommt mehr als einige Jahrhunderte zu spät. Die Menschen sind nicht mehr *unter sich*. Wir haben schon zu viele Handlungen an andere Aktanten delegiert, die nun unsere menschliche Existenz teilen« (Latour 2002: 231). Sinnvoll sei es, die Trennung von Subjekt und Objekt aufzugeben und im Modell des Netzwerkes zu denken. »Wer schießt, die Waffe oder der Mensch?«, fragt Latour und antwortet selbst: das Mensch-Waffen-Netzwerk (Krauss 2006: 435). Fahre *ich* mit meinem Auto langsamer, wenn Bodenschwellen verlegt sind? Der gesunde Menschenverstand wird diese Frage umstandslos bejahen. Latour erhebt Einspruch und erläutert, dass hier kein autonomes Subjekt handelt, sondern ein Kollektiv von Auto, Fahrer und Straßenschwelle. In diesem Kollektiv wirkt eine Reihe von Aktanten wie »verkehrspolitische Entscheidungen, juristische Normen, technische Konzepte von Ingenieuren und Ausführungen von Bauarbeitern, Baumaterialien, erwartbare Gewohnheiten von Autofahrern (Stoßdämpfer schonen), sensorisch unangenehme, physikalische Reaktionen des Autos in Abhängigkeit von Höhe der Schwelle und Geschwindigkeit« (Böhme 2006: 77; Latour 2002: 226–232). Von einem autonom handelnden *Ich*, das sein Fahrzeug beherrscht, kann aus dieser Perspektive keine Rede mehr sein.

Es mag provokant wirken, doch bei unseren Handlungen ist in hohem Maße von einer Mitwirkung der Dinge auszugehen. Unser

continues to evolve through us. (Miller 2005:8)

The notion that things do something to people was brought to a head by the French sociologist and philosopher Bruno Latour. That the human is lord over things that, indeed, humans themselves have created, is "a fine story, but it comes centuries too late. Humans are no longer *by themselves*" (Latour 1999: 190; emphasis in the original). It makes more sense to abandon the separation of subject and object and to think using the model of the network. "Who shoots, the weapon or the human?" Latour asks, before answering his question himself: the human-weapon network (Krauss 2006: 435, translation HH). Is it *me* who's driving my car slower when construction workers have put speed bumps into the road? Healthy common sense would answer this question with an unconditional yes. Latour raises objections however, elucidating that no autonomous subject is acting in this case, but rather a collective, consisting of the car, the driver and the speed bump. A variety of actants influence this collective, including "transport policy decisions, legal norms, technical concepts from engineers and tasks carried out by construction workers, building materials, predictable behaviours of car drivers (being careful with their bumpers), and sensorially unpleasant, physical reactions of the car, which depend on the height of the bump and on the car's speed" (Böhme 2006: 77; Latour 1999: 186–190; translation HH). From this perspective, it is no longer meaningful to talk of an "I" acting auton-

herkömmlicher Handlungsbegriff beruht auf einer verborgenen Asymmetrie, und zwar auf der Vorstellung, dass Menschen aktiv handeln und Dinge passiv sind, und diese möchte Latour offenlegen; er behauptet nun nicht, dass Dinge Subjekte mit Bewusstsein, Sprache und Intention seien. Er beharrt jedoch darauf, dass Dinge handlungsmächtig sind. Dinge entfalten als Aktanten Wirkung in komplexen Konfigurationen. Damit kritisiert er unser auf den Menschen konzentriertes Verständnis von Handlung und das allzu vertraute Denkschema, das feinsäuberlich zwischen sich gegenüberstehenden Polen wie Subjekt und Objekt oder Geist und Materie trennt. Gleichzeitig fordert er dazu auf, Objekte als soziale Akteure zu verstehen und sie in unseren Begriff von Gesellschaft mit einzubeziehen:

> Die Soziologen betrachten in ihrer Forschung eine größtenteils objektlose soziale Welt, auch wenn sie, wie wir alle, in ihrer täglichen Routine andauernd über die stete Gesellschaft, die fortwährende Intimität, die eingefleischte Zusammengehörigkeit, die leidenschaftlichen Affären, die verschlungenen Beziehungen der Primaten zu Objekten seit einer Million Jahre verwundert sein mögen. (Latour 2005: 82, Übersetzung PJBr)

omously, in exclusive control of their own vehicle.

Provocative as the assertion may seem, it is assumed that things influence our actions to a high degree. Our traditional concept of human action rests on a concealed asymmetry, on the notion that human behaviour is active while things are passive – this is the questionable axiom that Latour wishes to expose. He does not claim that things are subjects with consciousness, language and intention. Yet, he does insist that things possess agency. Effects unfold out of things, in their roles as actants, in complex configurations. In building this argument, Latour criticises an understanding of deeds and actions that concentrates too narrowly on humans and disputes an all too trusted way of thinking, which clinically separates opposite poles, whether these are conceived of as subject and object or as spirit and the material world. Concurrently, he encourages his readers to understand objects as social actors and to incorporate them into our concept of society:

> In their study, sociologists consider, for the most part, an object-less social world, even though in their daily routine they, like all of us, might be constantly puzzled by the constant companionship, the continuous intimacy, the inveterate contiguity, the passionate affairs, the convoluted attachments of primates with objects for the past one million years" (Latour 2005: 82).

Die Materialität von Flucht und Migration

Solche Ideen von Daniel Miller und Bruno Latour waren es, die uns dazu anregten, Migrationsforschung aus bislang ungewohnter Perspektive zu betreiben. Wie würde sich unsere Wahrnehmung von Migration verändern, wenn wir in den Spiegel der materiellen Welt blicken, Dingen unsere besondere Aufmerksamkeit schenken und konsequent die *Beziehungen* von Dingen und Menschen erforschen?

Wird die Gesellschaft von Dingen und Menschen akzeptiert und von einer Handlungsmacht nichtmenschlicher Akteure ausgegangen, stellen sich neue Fragen: Was berichten uns Dinge über menschliche Existenz, über Flucht, Rettung und Ankommen? Welche Funktionen haben Dinge im Rahmen von Flucht und Migration? Welche Versprechen tragen Dinge in sich, welche Emotionen, welche Aspirationen? Welche Dinge werden migrantischen Menschen als legitim zugesprochen, welche als illegitim abgesprochen? Welche Dinge entscheiden über den Status eines Flüchtenden als Schutzbedürftigen? Was sagt materieller Besitz über Menschenwürde und was bedeutet der Verlust von Hab und Gut für das Selbstbewusstsein und die persönliche Identität?

Das Miteinander von Menschen und Dingen tritt unter den Bedingungen von Flucht und Migration viel deutlicher zu Tage als in normalen Lebensabläufen. So ist das Packen eines Koffers oder Rucksacks eine Routinetätigkeit vor dem Antritt einer jeden Urlaubs- oder Geschäftsreise. Sind Menschen gezwungen, ihren Heimatort zu verlassen, ohne zu wissen,

The Materiality of Flight and Migration

Such ideas from Daniel Miller and Bruno Latour inspired us to conduct research into migration from perspectives that, to date, have not been habitual ones. How would our perceptions of migration change if we gazed into the mirror of the material world, gifted our special attention to things and were consequential in researching the *relations* between things and people?

If the concept of society as consisting of things and people is accepted, and if it is assumed that nonhuman actors possess agency, then new questions present themselves: What have things to report to us about human existence, about flight, being saved and arriving? Which functions do things have within contexts of flight and migration? Which promises do things carry with them, which emotions, which aspirations? Which things, categorized as legitimate, are granted to migrant persons, and which things are denied to them, categorized as illegitimate? Which things determine whether a refugee is awarded the status of a person requiring protection? What do material possessions tell us about human dignity, and what does the loss of worldly goods signify for a person's self-confidence and personal identity?

The togetherness of humans and things is exposed much more sharply to the light during the conditions imposed by flight and migration than it is in normal life-paths. Before departing for a holiday, or on a business trip, packing a suitcase or a backpack is a routine activity. But when

ob und wann sie zurückkehren können, wird die Auswahl der Gegenstände zu einer existenziell bedeutsamen und hoch emotionalen Handlung. Welche Kleidungsstücke schützen vor Regen, Hitze, Kälte? Welches Ding hilft womöglich in bedrohlichen Situationen, welches ist notfalls zu Geld zu machen? Welcher Gegenstand ist emotional unverzichtbar, weil er Gefühle zu geliebten Menschen weckt? Wie trenne ich mich von wertvollen Dingen? Was geschieht, wenn ich die falschen Dinge einpacke?

Verhandelt werden mit jedem Handgriff Spannungen und Widersprüchlichkeiten. Persönliche Geschichte(n), Hoffnungen, Abenteuer, schwer zu beschreibende Gefühle, Vergangenheit, Gegenwart und Zukunft werden in und mit Dingen transportiert (Löfgren 2016: 150; siehe hierzu auch den Beitrag von Veronika Reidinger und Anne Unterwurzacher in diesem Band).

Die alltägliche Routine des sich Ankleidens und des Schuheanziehens ist nicht weiter bemerkenswert. Unterwegs auf einer Fluchtroute hat das morgendliche Schnüren der Schuhe jedoch eine völlig andere Qualität. Gutes Schuhwerk ist für die Fortbewegung unerlässlich, denn blutige und schmerzende Füße machen das Fortkommen womöglich zunichte. Wenn von Grenzsoldaten die Schnürsenkel beschlagnahmt werden, bedeutet dies bedrohlichen Stillstand oder gar Scheitern. Dinge wie Schuhe und Schnürsenkel sind von elementarer Wichtigkeit, weil sie körpernah und in ihrer Funktion unverzichtbar sind (siehe den Beitrag von Anoush Masoudi in diesem Band). Neben den Dingen, die für nahezu alle Menschen essenziell sind, beeinflussen Faktoren wie Alter und Gender

humans are forced to leave the places they call home without knowing when or whether they will be able to return, the selection of objects turns into an existentially meaningful and highly emotional action. Which pieces of clothing protect against rain, heat, the cold? Which thing might help in threatening situations, which thing might have to be cashed in? Which object is impossible to do without, because it wakes feelings of loved persons? How can I part from valuable things? What happens if I pack the wrong things?

Each time a hand takes hold of something during this kind of packing, it negotiates tensions and contradictions. Personal history and histories, hopes, adventure, feelings that are hard to describe, and past, present and future are transported in and alongside things (Löfgren 2016: 150). Further insights into baggage and migration are recounted by Veronika Reidinger and Anne Unterwurzacher in this volume.

The everyday routine of getting dressed and of putting on one's shoes is usually unremarkable. But on routes people are using to flee, tying one's shoelaces in the morning takes on an utterly different quality. Good footwear is vital for making progress, while bleeding or aching feet can worsen someone's chances of moving forward. The taking of shoelaces by border troops means a perilous stoppage or even complete failure. Things such as shoes and shoelaces are of elementary importance, because they are physically close to bodies and functionally indispensable – see also Anoush Masoudi's article on shoes in this volume. And alongside the things that

die materiellen Bedürfnisse. Menstruierende sind während der Flucht und Migration auf den Zugang zu Menstruationsprodukten wie Binden, Tampons oder Menstruationstassen angewiesen und benötigen geschützte Orte für das Wechseln oder auch Reinigen dieser (siehe den Beitrag von Antonie Fuhse und Maliheh Bayat Tork in diesem Band).

Der tägliche Griff zum Smartphone ist für uns selbstverständlich. Unter den Bedingungen von Flucht und Migration wird das Smartphone zu einem überlebenswichtigen Gegenstand, und der Nachrichtenaustausch bekommt neue Relevanz. Das Gerät ermöglicht die Orientierung auf fremden Wegen und in unbekannten Orten. Es dient dem Geldtransfer, ist Wörterbuch, Radio und Fernseher, überdies Bild- und Tonarchiv der persönlichen Geschichte. Es verbindet Flüchtende untereinander und diese mit Angehörigen in der Ferne (siehe den Beitrag von Andrea Lauser, Miriam Kuhnke, Maliheh Bayat Tork und Antonie Fuhse in diesem Band). Der so ermöglichte Kontakt zu nahestehenden Menschen ist weit mehr als ein Austausch von Informationen. Die Gegenwart der Stimmen und Gesichter von Eltern und Geschwistern auf einem Smartphone-Bildschirm vermag es, starke Gefühle zu wecken. Trauer, Tränen, Trost, Angst, Verzweiflung, Freude und Wut werden über diesen handlichen Gegenstand transportiert.

Das Beispiel Smartphone macht darauf aufmerksam, dass wir mit Dingen nicht nur aufgrund ihrer Zweckmäßigkeit hantieren. Dinge haben vielfach eine affektive Dimension. Die englische Wendung »being in touch« bedeutet nicht nur »in Kontakt sein«, sondern – wörtlich übersetzt – auch

are essential for nearly all people, factors including age and gender have a major impact on people's material needs. Menstruating persons depend during flight and migration on menstruation products such as menstrual pads, tampons or menstrual cups and require protected places for changing or cleaning the same – see the essay by Antonie Fuhse and Maliheh Bayat Tork in this volume.

Reaching daily for our smartphones is something most of us take for granted. But under the conditions imposed by flight and migration, the smartphone becomes an object crucial to survival, and swapping news takes on a new relevance. The device provides orientation on strange paths and in unfamiliar localities. It executes money transfers, functions as a dictionary, radio and TV, and acts as an image and sound archive for personal stories. It connects refugees with each other and links people in flight to their family members, over great distances (see also the essay by Andrea Lauser, Miriam Kuhnke, Maliheh Bayat Tork and Antonie Fuhse in this volume). The communication facilitated to people who are psychologically close to the smartphone's owner goes far beyond a mere exchange of information. The presence of the voices and faces of parents and siblings on a smartphone screen can arouse strong feelings. Grief, tears, comfort, fear, desperation, joy and rage are transported via this wieldy object.

The example of the smartphone throws light on our dealings with things, interactions that are not grounded solely in the things' expediency. Things also fre-

»berührt sein« und verweist damit auf eine körperlich sinnenhafte Dimension, die zwischenmenschliche Kommunikation immer begleitet (Runia 2006: 5). Wir berühren Dinge, und Dinge berühren uns. Dinge sind dabei Medien von Gefühlen und befördern damit Selbstvergewisserung über Körperempfindungen. Das Schmuckstück, dessen Präsenz wir über Hautkontakt erfahren, die Fotografie, auf der uns die Blicke der Freunde berühren, das Plüschtier, das innig geherzt wird – solche Dinge sind Erinnerungs- und Gefühlsspeicher, denen im Verlauf von Flucht und Migration besonderer Wert zugeschrieben wird (siehe hierzu die Beiträge von Peter J. Bräunlein, Özlem Savaş und Elza Czarnowski in diesem Band).

Sich an etwas zu erinnern ist mehr als ein rein kognitiver Vorgang; körperlich sensorische und emotionale Prozesse sind mitbeteiligt. Besonders relevant ist dieser Zusammenhang, wenn es um das Essen geht. Die Koppelung bestimmter Speisen an Gefühle von Zugehörigkeit, Familie und Geborgenheit erfolgt über den Geschmackssinn. Heimat wird über Geschmack, Geruch und Gedächtnis erfahrbar und der direkteste Weg in die emotionale Heimat führt über das Zubereiten und Verspeisen von Gerichten, die seit früher Kindheit vertraut sind. Dieser Vorgang des gemeinschaftlichen Essens, die Mahlzeit, ist ein sinnliches und soziales Ereignis. Eine Tischgemeinschaft bindet und verbindet, bestätigt und erweitert Familienstrukturen und festigt Freundschaftsbande (siehe hierzu den Beitrag von Friedemann Yi-Neumann in diesem Band).

Ding-Mensch-Beziehungen entfalten also auf unterschiedlichen Ebenen ihre Bedeut-

quently have an affective dimension. The English phrase "being in touch" does not merely connote being "in contact", but also "being touched", and thus points to the bodily and sensory dimension that always accompanies communication between people (Runia 2006: 5). We touch things and things touch us. In this sense, things are mediators of feelings, and as such they promote self-assurance about our bodily sensations. The piece of jewellery, whose presence we experience via our skin, the photograph on which our friends' gazes touch us, the cuddly toy that is hugged fervently – such things are reservoirs of memories and feelings, to which particular value is attributed in the course of flight and migration – as is explored by the contributions from Peter J. Bräunlein, Özlem Savaş and Elza Czarnowski to this volume.

Remembering something specific is more than a purely cognitive process; bodily, sensory and emotional processes also play their parts. These interconnections are especially important when the subject in question is food. Through the sense of taste specific dishes are coupled together with feelings of belonging, family and emotional security. People experience home through the channels of taste, smell and memory, and the most direct route to an individual's "emotional home" runs via the preparation and eating of dishes familiar since early childhood. The meal, this process of communal eating, is a sensory and social event. A "community of the table" binds and connects, confirms and extends family structures, and reinforces ties of friendship – Friedemann Yi-Neumann

samkeit für Flucht und Migration. Mit im Spiel sind Körper, Sinneswahrnehmung, Funktion, Emotion, Erinnerung, Gemeinschaftsbildung und Kommunikation. Wenn wir Mensch-Ding-Beziehungen unter solchen Gesichtspunkten erforschen, sind damit die »großen« Themen von Flucht und Migration nicht ausgeblendet. Fluchtgründe, strukturelle Ungleichheit, Rassismus, Grenzgewalt, Menschenrechtsverletzungen, Staatenlosigkeit, Bürokratie und Asylpolitiken – auch dies wird im materiellen Spiegel von Flucht und Migration reflektiert. Allerdings sind solche Themen nicht abstrakt vorzufinden, sondern nahe an Menschen, in Dingen und deren Geschichten. Das Erzählen ist das methodische Prinzip unseres Forschungsansatzes und die exemplarische Geschichte ihr Medium. Konsequenterweise war unser Ehrgeiz darauf gerichtet, Dinge und Geschichten von Menschen und ihren Dingen zu sammeln. Dinge, so zeigt sich, können Verbindungen herstellen über soziale, nationale oder kulturelle Grenzen hinweg. Dies zeigt sich immer wieder beim gegenseitigen Geben und Nehmen [gegenseitigen Gaben], wie Ethnolog:innen vielfach gezeigt haben (Mauss 2019). Dinge vermögen allerdings auch zu trennen, indem sie als Hindernis im Weg stehen (wie zum Beispiel Grenzzaun und Stacheldraht) oder Menschen markieren und identifizieren (wie zum Beispiel der Fingerabdruck oder die Formulare der Asylbehörden). Dinge, denen wir im Kontext von Flucht und Migration begegnen, haben bestimmte Funktionen, und gleichzeitig fällt die Wandelbarkeit dieser Funktionen ins Auge. Ein Reisepass ist in den meisten Situationen ein wichtiger Türöffner, in anderen Situationen kann sein Besitz

engages more deeply with food, home and memory in this volume.

Thing-people relationships evidently reveal their significance for flight and migration on different emotional and theoretical levels. Consistent elements in these equations are bodies, sensory perceptions, function, emotion, memory, the building of community and communication. Researching human-thing relationships primarily through these aspects does not mean blanking out the "big themes" of flight and migration. Motivations for fleeing, structural inequality, racism, border violence, human rights violations, statelessness, bureaucracy, and asylum policies and politics – these issues are also reflected in the material mirror of flight and migrations. That said, such themes are not discussed in an abstract manner, but remain close to people, and located in things and their stories. Indeed, story-telling is the methodological principle underpinning our research approach, and the characteristic story is its medium. Consequently, our ambition was focussed on collecting things – and stories – about humans and their things. Things can establish connections across and beyond social, national and cultural borders.

This is shown, for example, in the reciprocal giving and receiving of gifts, as anthropologists have demonstrated on numerous occasions (Mauss 2019). Things, however, are also capable of dividing, by blocking the path – border fences and barbed wire, for example – and of marking and identifying people – as is the case of fingerprints or the forms archived in asylum authority offices. Things we encounter

höchst gefährlich sein (siehe hierzu Romm Lewkowiczs Beitrag in diesem Band). Wenn wir also Dinge und Geschichten sammeln, ist damit das Bemühen um Empathie und die Überzeugung verbunden, dass hier ein Zugang zum Verstehen komplexer Migrationsverhältnisse eröffnet wird (siehe hierzu den Beitrag von Nina de la Chevallerie in Kooperation mit dem »boat people projekt« in diesem Band).

Wenn es in diesem Band um »Moving Things« geht, dann genau im doppelten Sinne des Begriffs (Basu & Coleman 2008: 317): Dinge werden präsentiert, die in räumlicher Bewegung waren und von dieser Bewegung erzählen. Und gleichzeitig handelt es sich um Dinge, die gefühlsmäßig bewegen, und zwar jene Menschen, die eine Beziehung zu ihnen hatten (und haben), aber auch andere, die sich bewegen lassen. Zum Dritten setzen wir selbst »die Dinge« in Bewegung, indem wir sie unter verschiedenen Gesichtspunkten betrachten, analysieren und ausstellen.

in the context of flight and migration have particular functions, and yet observers are also struck by their mutability. In most situations, a passport is an important document for opening doors, but in others, possessing such an artefact can be highly dangerous; see Romm Lewkowicz' essay in this volume. Thus, in collecting things and stories, the effort to practice empathy is tied to the conviction that this practice widens access to the understanding of the complex circumstances of migration. Such questions are investigated in the article by Nina de la Chevallerie in cooperation with the "boat people projekt" in this volume.

If this volume is indeed about *Moving Things*, then precisely in a double sense (Basu & Coleman 2008: 317): Things are presented that have moved spatially and that also narrate the story of this movement. Concurrently, the matter at hand are things that move people's feelings and, more concretely, move those people who had or have a relationship to them – but also others who allow themselves to be moved. Finally, we ourselves release "the things" into movement, by observing their different aspects, by analysing them and by exhibiting them.

Literatur | References

Basu, Paul; Simon Coleman. 2008. Introduction: Migrant Worlds, Material Cultures. *Mobilities* 3 (3), 313–330.

Böhme, Hartmut. 2006. *Fetischismus und Kultur. Eine andere Theorie der Moderne.* Reinbek.

Krauss, Werner. 2006. Bruno Latour: Making Things Public. In: Stephan Moebius, Dirk Quadflieg: *Kultur. Theorien der Gegenwart.*, 430–444. Wiesbaden.

Latour, Bruno. 2002. *Die Hoffnung der Pandora: Untersuchungen zur Wirklichkeit der Wissenschaft.* Frankfurt.

Latour, Bruno. 2005. *Reassembling the Social: An Introduction to Actor-Network Theory.* Oxford.

Löfgren, Orvar. 2016. Emotional Baggage. Unpacking the Suitcase. In: Jonas Frykman, Maja Povrzanović Frykman: *Sensitive Objects. Affect and Material Culture*, 125–152. Lund.

Mauss, Marcel. 2019. *Die Gabe. Form und Funktion des Austauschs in archaischen Gesellschaften.* Frankfurt a.M.

Miller, Daniel. 2005. Introduction. In: Daniel Miller: *Materiality*, 1–50. Durham.

Miller, Daniel. 2008. Material Culture. In: Tony Bennett, John Frow: *The Sage Handbook of Cultural Analysis*, 271–290. London.

Runia, Eelco. 2006. Presence. *History and Theory* 45, 1–29.

Anmerkungen

1 Dies ist eine gekürzte und überarbeitete Version der Geschichte »Schnürsenkel, Gürtel, GPS, Google Maps: Die wesentlichen Dinge Flüchtender«. Die komplette Geschichte findet sich auf der Website des Forschungsprojektes (Maliheh Bayat Tork: »Schnürsenkel, Gürtel, GPS, Google Maps: Die wesentlichen Dinge Flüchtender«, *Zur Materialität von Flucht und Migration*, abgerufen am 29.10.2021. https://materialitaet-migration.de/objekte/shoe-lace-belt-gps-google-map-vital-things-to-the-people-on-the-run/).

2 Das Forschungsprojekt wird von drei Verbundpartnern getragen: dem Institut für Ethnologie der Georg-August-Universität Göttingen, dem Museum Friedland und dem Berliner Ausstellungsbüro »Die Exponauten. Ausstellungen et cetera«. Zum Forschungsprojekt siehe https://materialitaet-migration.de/

Notes

1 This is an abridged and revised version of the story "Shoe Laces, Belt, GPS, Google Maps: Vital Things to the People on the Run". The complete story can be followed on the research project's website (Maliheh Bayat Tork: "Shoe Laces, Belt, GPS, Google Maps: Vital Things to the People on the Run", *Zur Materialität von Flucht und Migration,* accessed 29 October, 2021. https://materialitaet-migration.de/en/objekte/shoe-lace-belt-gps-google-map-vital-things-to-the-people-on-the-run/

2 The research project was facilitated by a cooperation of three institutions: the Institute for Social and Cultural Anthropology, at the University of Göttingen, the Friedland Museum and the Berlin-based exhibition agency "Die Exponauten. Ausstellungen et cetera". For more on the research project, see: https://materialitaet-migration.de/en/

»Moving Things« erforschen
Ein Gespräch des Projektteams
über Herausforderungen und Chancen

Researching "Moving Things"
A Project Team Discussion
About Challenges and Chances

Maliheh Bayat Tork | Antonie Fuhse | Andrea Lauser | Friedemann Yi-Neumann | Peter J. Bräunlein | Joachim Baur

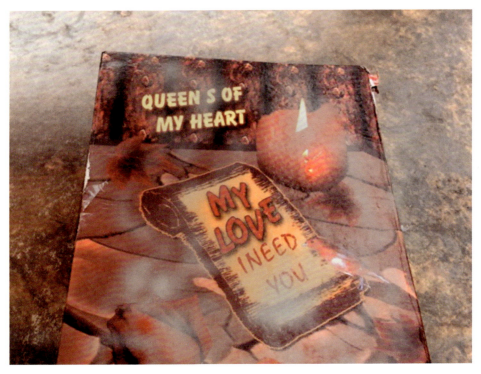

Abb. 1: Mit Staub bedeckte Schachtel von Um-Abdallah, 2019. ©Maliheh Bayat Tork

Image 1: Box belonging to Um-Abdallah covered with dust, 2019. ©Maliheh Bayat Tork

Friedemann: Was war der eindrucksvollste Gegenstand, auf den du gestoßen bist oder die beeindruckendste Erfahrung, die du während der Forschung im MatMig-Projekt[1] gemacht hast? Was hat dich überrascht?

Maliheh: Ich kann gar nicht sagen, welches der Objekte mich am meisten beeindruckt hat. Als ich mit meiner Beschäftigung im Projekt angefangen habe, redete ich mit vielen Menschen. Sie zeigten mir zum Beispiel ein Schmuckstück, ein Erinnerungsobjekt von Zuhause oder etwas, das ihre Eltern oder andere Familienmitglieder ihnen gegeben hatten. Die Art und Weise, wie sie über ein solches Ding redeten, wie wichtig es ihnen war, wie sie sich anstrengten, um es behalten und mitnehmen zu können, machte mir bewusst, wie stark die Beziehungen zwischen Menschen und Dingen sind. Aber für mich gibt es nicht das eine Ding. Es gab beeindruckende Geschichten, wie die von Um-Abdallah, einer Syrerin, die über eine Schachtel sprach, die sie als Geschenk zum Muttertag bekommen hatte. Um-Abdallah hatte die Schachtel aus den Trümmern und der Asche ihres zerstörten Hauses in Syrien gerettet. Den Staub auf der Schachtel entfernte sie aber nicht. Dieser Moment – als wir über die Schachtel, die sie über die große Entfernung mitgebracht hatte, redeten und diese genau betrachteten – war sehr rührend. Ich war sehr vorsichtig, um nicht den Staub auf der Schachtel zu entfernen, der sie an all das erinnerte, was sie durchgemacht hatte. Letztendlich ist es der Forschungsprozess, der mich beeindruckte, dass wir Beziehungen und Verbindungen, die Menschen zu ihren Dingen aufgebaut haben, untersuchten und damit ihre Geschichten erzählten.

Andrea: Mich hat bei all den Themen des Forschungsprojekts schon immer die ganz exis-

Friedemann: What was the most impressive thing – experience or object – you met with during your research in the MatMig[1] Research Project? The thing that surprised you the most?

Maliheh: I can't actually say which of the objects impressed me the most. As I engaged more deeply in the project, I started talking to many people. Some showed me, for example, a piece of jewellery, a keepsake from home or something their fathers or mothers or other family members had given them. The way they talked about these objects, the way they held them dearly, the way they struggled to keep them made me realize how strong the ties are between people and things. There is not one specific thing that stands out for me in the research. But there were impressive stories, like the one from Um-Abdallah, the Syrian woman who talked about a box that was a Mother's Day gift. Um-Abdallah had saved this box from among the rubble and ashes of her burned-out house in Syria. She did not even want to clean away the dust on this box. This moment was very touching – when we talked about it and examined together this gift box she had carried all the way. I was so careful not to destroy the dust on the box that reminded her of all she had been through. Ultimately, it was the research process itself that also impressed me, how we investigated relations and connections that people have built up to their things – and in this way also narrated their stories.

Andrea: During the work on all these research project themes, an utterly existential question continued to move me on a

tenzielle Frage persönlich berührt: Was wären für mich Dinge, die ich unbedingt dabeihaben müsste, besitzen müsste? Peter und ich haben ja fast zwei Jahre in einer Gesellschaft gelebt, bei den Mangyan auf der Insel Mindoro (Philippinen),[2] deren materielle Kultur recht überschaubar ist, ja man könnte sie auch als minimalistisch bezeichnen. Dort gibt es überdies heftige Taifune, die zyklisch gravierende Zerstörungen hinterlassen. Sich nicht an Dinge zu binden, das habe ich als pragmatische Anpassungsfähigkeit interpretiert – als eine Art materieller Minimalismus. Doch im Kontext von Krieg, Flucht und Vertreibung haben Zerstörungen und Besitzverlust sicherlich eine andere Dimension, die für Menschen oft den Verlust oder die starke Veränderung ihres gewohnten Lebens bedeutet.

Maliheh: Ich denke auch, dass die Frage nach Besitz eine existenzielle ist. Viele Migrant:innen müssen sich an eine neue Umgebung anpassen und lernen, dass Dinge kommen und gehen, damit sie nicht zu sehr über den Verlust von Dingen trauern. Genau das beobachtete ich in den Lagern für Geflüchtete im Iran und in Deutschland, und dies ist auch etwas, das ich selbst als Migrantin lernen musste. Früher, während meiner ersten Zeit hier in Deutschland, weigerte ich mich, viele Dinge zu kaufen oder zu sammeln. Ich hatte Angst davor, mich in die Sachen zu verlieben, und fragte mich, was passieren würde, wenn ich in den Iran zurückginge: Würde ich traurig sein? Das war eine Wirkung, die mein temporäres Leben hier in Deutschland auf mich hatte. Nach und nach habe ich aber gelernt, anders damit umzugehen, die Dinge zu schätzen und mich aber auch von ihnen emotional zu distanzieren, wenn ich sie zurücklassen muss.

personal level: What were for me the things that I simply had to have on my person or otherwise possess? Peter and I lived for almost two years in a society – the Mangyan society on the island of Mindoro in the Philippines[2] – whose material culture is rather modest, or you could even say minimalistic. The island is regularly hit by fierce typhoons, which leave serious destruction behind them on a cyclical basis. I interpreted the Mangyan's way of not attaching themselves to things as pragmatic adaptability – as a kind of material minimalism. But in the context of war, flight and expulsion, destruction and the loss of property harbour an unquestionably different dimension, which often signifies for individuals the loss of, or at least a massive change in, their normal way of life.

Maliheh: I agree that this question of possession is an existential one. Many migrants must adapt to new environments and to learn that things come and go, so that they do not grieve over losing things too much. This is what I observed in the refugee camps in Iran and in Germany and what I personally learned as a migrant myself. Formerly, here in Germany, I refused to buy or collect much stuff, because I had the fear: What if I fall in love with these things and then, when I go back to Iran, I cannot take them with me and will be sad? This was one of the effects of life, experienced as something temporary, here in Germany. But bit by bit I've learned to function differently, to appreciate things and yet still to be able to emotionally distance myself from them, whenever I have to leave them behind.

Antonie: I notice that I also don't have a

Antonie: Ich merke, dass ich auch kein Lieblingsobjekt habe, aber ich finde die Objekte am interessantesten, die so ganz alltäglich daherkommen oder erstmal unscheinbar sind und doch ganz viele praktische Seiten an sich haben. Zum Beispiel dieses Stück Stoff von Wael im Katalog unserer Projektwebseite,[3] das mal zu einer Decke gehört hat. In dem Text wird gezeigt, wie viele unterschiedliche Funktionen eigentlich so eine Decke erfüllen kann: Sie kann auf der Flucht als Decke verwendet werden, wenn nichts anderes zur Verfügung steht, als Unterlage auf einem Bett, als Teppich, als Vorhang. Je nachdem, was gerade gebraucht wird, wechselt dieses Stück Stoff die Form und die Bedeutung. Es beinhaltet aber immer auch eine symbolische Ebene, weil die Decke einst diesem jungen geflüchteten Mann von seiner Großmutter geschenkt wurde. Deswegen ist es auch ein Erinnerungs-

favourite object amongst those we have worked with, but the objects I find most interesting are those that seem to be entirely mundane, or at least unremarkable at first, but that have many practical sides to them nonetheless. This piece of cloth from Wael in the catalogue of our project website,[3] for example, which once belonged to a blanket. The text demonstrates how many different functions a blanket can fulfil: as a blanket to sleep under during flight, when nothing else is available; as a bed sheet on a bed; as a carpet; as a curtain. Depending on what is needed in a particular instant, the significance and form of this piece of fabric are transformed. But it continues to contain the symbolic dimension, because the blanket was once given to this young, male refugee by his grandmother. This also makes it a memento, which is also why it was import-

Abb. 2: Das Stück Stoff von Wael, 2019. ©Samah Al-Jundi Pfaff und Katharina Brunner

Image 2: Wael's piece of cloth, 2019. ©Samah Al-Jundi Pfaff and Katharina Brunner

objekt, weswegen es ihm auch wichtig war, ein Stück davon zu behalten, als er auf der Flucht nicht viel mitnehmen konnte. Dennoch beinhalten viele solche oder so ähnliche Dinge etwas ganz Praktisches.

Friedemann: Das Spannende an diesem Objekt ist auch – in der Wissenschaft wird ja immer über Transformation geredet, ein großer und recht abstrakter Begriff, dass dieses Stück Stoff einen Transformationsprozess im Zuge einer Fluchtgeschichte an einem Ding, seinem materiellen Wandel und seiner Verwendung handfest greifbar macht. Es zeigt, wie sich Dinge durch Migration verändern.

Antonie: Genau aus diesem Grund ist jetzt auch das Kapitel von Maliheh und mir zum Thema Menstruationsprodukte in diesem Begleitband. Das sind ja ganz alltägliche Produkte – zumindest für Frauen –, und das kommt manchmal zu kurz, was für praktische Themen beim Fliehen und Migrieren auf einen zukommen, was man mitnimmt, etwa Tampons und Binden. Aber auf der anderen Seite zeigt es auch, dass die Infrastruktur da sein muss, wie Toiletten, um Menschen in Camps vor Ort zu versorgen. Diese Dinge haben mich selbst immer wieder zum Nachdenken angeregt, etwa darüber, wie viele praktische Dinge ich habe und dass ich mir überhaupt gar keine Gedanken darüber machen muss. Dinge, die ganz selbstverständlich für mich sind. Aber in Fluchtkontexten oder dem Leben in Camps sind diese Dinge eben nicht selbstverständlich.

Peter: Es ist eine Fotografie, die mich zum Nachdenken anregte, und zwar vor allem über den Zeithorizont, in dem wir unsere Forschung durchgeführt haben. Wir hätten eine andere Forschung gemacht, wenn wir in den

ant for him to keep just a piece of it, even when he knew he couldn't take much with him when fleeing. And yet, irrespective of this symbolic and emotional significance, such and similar objects retain something entirely practical.

Friedemann: Another exciting thing about this object – in an age in which scholarly discourse is constantly about transformation, a rather large and abstract concept – is that this piece of fabric makes tangible a transformation process, during a story about flight, through the presence of a single thing and its material metamorphosis and utilization. It shows how things change through migration.

Antonie: It is for precisely this reason that the chapter by Maliheh and myself on menstruation products is included in this volume. These are utterly everyday products, at least for women, and what practical issues people must face up to when fleeing and migrating are sometimes not talked about enough: What do individuals take with them? Tampons and menstrual pads, for example.

But on the other hand, this issue also demonstrates that the infrastructure has to be there, toilets, for example, as part of the on-the-ground provision for people in camps. Such things have often prompted me to reflect about how many practical things I have, for example, and about the fact that I don't have to give a thought to them at all. These are things that I take entirely for granted. But in the context of flight, or of life in the camps, women and others cannot take such things for granted.

Peter: It's one particular photo that gets

60er-, 70er-, 80er- oder 90er-Jahren geforscht hätten. Wir sind heute umgeben von dramatischen Bildern, wie etwa von Menschen in überfüllten Lagern oder in Schlauchbooten auf dem Mittelmeer. Mich hat ein Foto sehr bewegt, das ich auf einer Zeitschrift mit dem Titel *Zeithistorische Forschungen* gefunden habe.[4] Da geht es um Flucht und Migration im Grenzdurchgangslager Friedland bei Göttingen. Auf dem Foto sind vietnamesische Geflüchtete zu sehen, die 1978 in Friedland Aufnahme gefunden haben. In einem Kinderwagen befindet sich ein Umzugskarton, auf dem steht »Umzug ist Vertrauenssache«. Boatpeople aus Vietnam finden also Aufnahme im Nadelöhr Friedland und werden empfangen von einer Rot-Kreuz-Schwester, die vermittelt »Alles ist gut«, »Vertraut uns, wir stehen an Eurer Seite«. Dann sehe ich ein triviales Objekt, den Umzugskarton. Der hat für mich eine starke Symbolkraft. Ich denke an eigene Umzüge und das Packen von Kisten und die damit einhergehende Ortsveränderung. Umzüge sind eine existenzielle Sache. Dieser im Bild erfasste »Umzug« hat allerdings eine ungleich dramatischere Dimension. Er erinnert nicht zuletzt daran, dass Gesellschaften bereit waren, bestimmten geflüchteten Familien, die vom Tod bedroht waren, zu helfen. Und das erweckt in mir den Wunsch, über die historische Dimension von Flucht zu sprechen. Eine Dimension, die eben auch zeigt, dass Hilfe und Zuwendung für ausländische Geflüchtete damals staatlich gewünscht und verordnet waren. In der Nachkriegszeit war das einmalig. Gleichzeitig existierte damals wie heute große Skepsis, wir sind umzingelt von populistischen Stimmen, von Ängsten, Befürchtungen, Abwehrreaktionen. »Umzug

me thinking again and again, with my thoughts going beyond the current timeframe in which we carried out our research. Our study would have been substantially different if we'd been researching in the 1960s, 1970s, 1980s, or 1990s. We are surrounded today by dramatic images of people in overcrowded camps or in rubber dinghies on the Mediterranean. I was very moved by a photo I found on the cover of a magazine called *Zeithistorische Forschungen/Studies in Contemporary History*.[4] The topic addressed is the Friedland Border Transit Camp near Göttingen. The photo shows Vietnamese refugees who were housed in Friedland in 1978.

A pram is loaded with a moving box, bearing the advertising slogan "Moving is a Matter of Trust" [*"Umzug ist Vertrauenssache"*]. We thus see Vietnamese Boat People being granted a place to live in Friedland – which functions as the eye-of-the-needle to the life beyond – and being welcomed by a Red Cross worker who seems to communicate "Everything is OK", and "Trust us, we're on your side." I then see the trivial object, the moving box, which possesses a strong symbolic power for me. I think of my own moves, the packing of boxes and the changing of places that entails. Moving is an existential matter. Not the least, the photo provides a testimony that societies were prepared to help specific refugee families who were in danger of dying.

And that makes me want to speak about the historical dimension of flight and migration. A dimension that demonstrates that, in the period in question, states want-

ist Vertrauenssache«: Dieser triviale Spruch auf einem Umzugskarton im Lager Friedland aus dem Jahr 1978 hat mir sehr zu denken gegeben, wie sich Aufnahmegesellschaften und Geflüchtete zueinander verhalten und wie unser Blick als Forschende davon beeinflusst wird.

Antonie: Du hast den Kontext in den Vordergrund gerückt. In welchem historischen Moment bewegen wir uns eigentlich? Die Diskurse, die wir jetzt gerade gesellschaftlich erleben, haben auch einen großen Einfluss darauf, wie wir die Forschung betreiben. Allerdings zeigt das Beispiel auch, finde ich, dass es schon seit Langem immer wieder Fluchtbewegungen gibt, und ich glaube, es

ed to provide help and care for foreign refugees and ordered that others in society follow suit. This was unique in postwar history. Concurrently, then as now, a great deal of scepticism existed: Now, as then, we are ringed in by populist voices, from anxieties, fears and defensive reactions. "Moving is a Matter of Trust": This trivial slogan on a moving box in the Friedland Camp in 1978 made me think a lot about relations between receiving societies and refugees, and how this influences our perspective as researchers.

Antonie: You have brought the context to the fore. Which begs the question: How to describe the historical moment we are

Abb. 3: Rot-Kreuz-Schwester mit Geflüchteten aus Vietnam im Durchgangslager Friedland, 1978. ©Helmuth Lohmann/Picture Alliance

Image 3: A Red Cross worker in 1978, surrounded by Vietnamese refugees at Friedland Transit Camp ©Helmuth Lohmann/Picture alliance

gab auch immer schon Gegendiskurse. Die Frage ist doch: Warum gibt es immer noch sofort Gegenbewegungen? Warum ist unser Migrations- oder Asylsystem so, wie es ist, und warum sind jetzige Fluchtbewegungen so, wie sie sind? In welche globalen und historischen Verhältnisse sind sie eingebettet? Im Moment prägt die Covid-19-Pandemie viele Aspekte unseres Lebens und ist neben anderen Krisen, wie der Klimakrise, zu einer entscheidenden globalen Herausforderung geworden. Neben den persönlichen sozialen und wirtschaftlichen Krisen, die sie ausgelöst hat, stellt die Pandemie die ethnologische Feldforschung vor neue Aufgaben. Maliheh, du hast mit Hilfe von Skype, Instagram & Co von der Forschung vor Ort zu einer Forschung aus der Ferne gewechselt. Welche Rolle spielen die Dinge in deiner Forschung aus der Ferne? Wie hat sich die Rolle der Dinge in der Forschung verändert?

Maliheh: Ich kann sagen, dass vieles in der Forschung gleich geblieben ist. Ich kann die Objekte zwar nicht anfassen, aber die Menschen zeigen mir Bilder ihrer Pässe, ihrer Ausweise, von Dokumenten oder von anderen Dingen, die sie ein- und umpacken. Wir teilen Dinge über Videoanrufe, aber natürlich kann ich diese Objekte nicht anfassen. Aber trotz dieser Einschränkungen konnte ich den Migrant:innen auf ihren Routen folgen. Ich konnte aus der Ferne beobachten, wie Dinge, zum Beispiel Kunstwerke, die die persönliche Entwicklung Migrierender reflektieren. Deswegen denke ich, dass es Methoden gibt, die das Potenzial haben, auch über die Ferne Gefühle zu vermitteln und die Dichte der Beziehungen zwischen Menschen und Dingen zu erfassen.

presently in? The discourses that we experience as a society also have a substantial influence on how we conduct research. That said, your example also reminds us, I find, that flight and migration have been part of human history for a long, long time, and I'm convinced that there have always been polemics and societal conversations aimed against refugees and migrants. But the question remains: Why do opposition movements still always emerge immediately? Why are our systems for migration and asylum the way they are, and why are current migrations the way they are? In which global and historical relations are they imbedded? At present, the Covid-19 pandemic is defining many aspects of our lives and has become a decisive global challenge alongside other crises, including the climate crisis. Parallel to the personal, social and economic crises that it has unleashed, the pandemic presents fresh challenges to ethnological field research. Maliheh, you have switched from research on the ground to distance-based research with the help of Skype, Instagram and other digital communication tools. What role do things themselves play in your distance-based research? How has the role of things in research altered?

Maliheh: I would say that much has remained the same. It's just that I can't touch the objects, but people show me pictures of their papers, ID-papers, documents or the things they have packed/unpacked, and we share things through video calls, but, of course, I cannot touch these objects. Even though the objects were only digitally accessible for me and I conducted

Friedemann: Ein anderes historisches Ereignis, das du, Peter, ja schon genannt hast, ist der lange Sommer der Migration 2015 und die damit verbundene humanitäre Krise. Die Perspektive unseres Forschungsprojektes ist durch dieses Ereignis mitgeprägt worden. Als ich 2015 angefangen habe, zu diesem Thema zu arbeiten, fanden über eine Million Menschen den Weg nach Europa und zunächst wurden viele ›willkommen geheißen‹. Dann passierte eine heftige Diskursverschiebung, es kam zu rassistischer Gewalt gegen Geflüchtete. Gleichzeitig verlagerte der EU-Türkei-Deal, d.h. die Schließung der EU-Außengrenzen, die humanitäre Krise vom Zentrum an die Ränder Europas. Was mir in Sachen Materialität und Fluchtmigration der letzten Zeit dazu einfällt, ist das Lager Moria auf Lesbos, das abbrannte. Das Lager Friedland, in dem wir geforscht haben, ist hingegen auf den ersten Blick eine relativ gut geordnete, beschauliche Infrastruktur. In Friedland spiegelte sich die Krise unscheinbarer, gerade in den Geschichten der abgelehnten Asylbewerber:innen, deren Hoffnungen auf eine sichere Zukunft in Deutschland nicht in Erfüllung gegangen sind.

Andrea: Die Dynamiken in der Krise, die du ansprichst, sind etwas, das mich auch aus einer sozialwissenschaftlichen Perspektive grundsätzlich interessiert. Es handelt sich um etwas, das als eine Art ›Kollapsologie‹ bezeichnet werden könnte.[5] Krisen und mehr oder wenige heftige, existenzielle Kollapse gab und gibt es immer wieder und wurden ja auch in der Ethnologie an so genannten ›klassischen Beispielen‹ erforscht.[6] Dabei wurde deutlich, dass Menschen zu Beginn einer Krise vor allem mit Solidarität reagieren. Aber dann, je länger die Krise anhält, folgen

the research remotely, I was able to follow migrants on their routes. I could remotely observe how things, like pieces of art, can reflect the personal development of migrating people over time. Thus, I think that methods of ethnography exist, which have the potential to mediate feelings across distances, and which can capture the density of relations between humans and things across the same.

Friedemann: A further historical event that you, Peter, have already mentioned is the long summer of migration in 2015 and the humanitarian crisis that accompanied it. This event was a major factor in shaping the perspective taken in our research project. While I started working on this theme in 2015, over a million people were making their way to Europe, and at first many were "welcomed". But then public discourse about the refugees became substantially distorted, a shift that was followed by racist violence against the latter. Concurrently, the humanitarian crisis was shunted from Europe's geographical centre to its margins by the deal between the EU and Turkey. This effectively meant the EU was closing its outer borders. What then comes into my mind when I think about materiality and refugee migration in more recent years is the Moria Refugee Camp on Lesbos, which burnt to the ground. By contrast, the Friedland Camp, in which we did our research, is a relatively well-ordered and tranquil piece of infrastructure – at least when viewed superficially. In Friedland, the general crisis is reflected in less visible ways in the lives of its inhabitants: in the stories of asylum-seekers who have been refused asylum, for exam-

Phasen mit unsolidarischen, egozentrischen, abweisenden und abwertenden Reaktionen. Eine solche Abfolge von Phasen ist sozialwissenschaftlich in vielen Fällen belegt und mit dieser »Kollapsgeschichte« müssen wir uns auseinandersetzen. Wie gehen wir damit um, auch um das Humane zu retten und zu bewahren, nicht nur in der Anfangsphase, sondern es auch transformativ durchzuhalten und zu leben?

Peter: Hier kommt unser Ansatz zum Tragen, der bewusst die Mikroperspektive, das Erzählen und Erinnern, in den Mittelpunkt stellt. Wir beschäftigen uns mit sozialen Beziehungen der Menschen zu ihren Dingen und der Menschen untereinander. Letztlich knüpfen wir, und das ist die Stärke unseres Ansatzes, finde ich, an die Tradition des Geschichtenerzählens an. Das ist sehr lebendig, vielfach berührend und lässt sich doch schwer systematisieren. Wir sagen nicht, Migration ist so und so, sondern: Hört euch die Geschichten an! Das ist der Wert unserer Forschung.

Friedemann: Vielleicht liegt ja auch in dieser Art der Forschung eine gewisse Antwort auf Andreas Frage, die sie bezüglich der langfristigen Solidarität gestellt hat: jenseits der Events empathisch und reflektiert über Migration zu schreiben. Die Fähigkeit der Dinge in der Ethnografie, die persönlichen Geschichten, aber eben auch die weiteren strukturellen, gesellschaftlichen und materiellen Zusammenhänge zu reflektieren, halte ich ebenso für grundlegend. Wenn das in einem guten Verhältnis zueinander steht, dann ist für mich die (dingorientierte) Ethnografie geglückt.

Peter: In der Einleitung zu diesem Band habe ich ja das Beispiel des Kofferpackens ange-

ple, and whose hopes for a secure future in Germany have not been fulfilled.

Andrea: The crisis dynamics you address are something that interest me fundamentally from a social-scientific perspective. This is about something that could be termed a kind of "Collapsology".[5] Crises and existential collapses, sometimes more and sometimes less vehement by nature, are things that have always reoccurred and are still reoccurring, and that are researched in social anthropology by applying so-called "classic examples".[6] These make evident that people's reactions at the start of a crisis are primarily ones of solidarity. People demonstrate solidarity spontaneously. But the longer the crisis goes on, this initial phase of solidarity is followed by phases of antisolidarity and of actions characterised by egocentrism, rejection and disparagement. The social sciences have provided evidence for just such a series of phases in many different instances of crisis: The story and history of collapse is one that people will continue to be confronted with. How do we deal with the experience of these phases, save and protect what is humane in society? And how to persevere transformatively with what is humane, not merely in the initial phase of a crisis, and embody the humane in our lives?

Peter: Here is where our approach comes into its own, which consciously places the microperspective and the acts of narration and remembering at the centre of our activities. We study the social relations of people to their things and of people to each other. Finally, we draw on the tradition of storytelling, which I see as the

dacht, um den existenziellen Moment des Auswählens zu illustrieren. Ich war angeregt durch Jason de Leóns Text[7] über seine Begegnung mit Migrant:innen an der Grenze zwischen Mexiko und den USA, die genau solche Überlegungen anstellen mussten: Was nehme ich mit? Was ist überlebensnotwendig, um durch die lebensfeindliche Wüste Arizonas zu kommen, die Grenze der USA zu überqueren und den Häschern der Grenzpolizei zu entkommen? Aber auch: Welches Ding ist erinnerungswürdig?

Joachim Baur: Und das sind zweifellos entscheidende, einschneidende Fragen. Aber ich habe mich und dich dabei gefragt, ob dieser Moment des Kofferpackens nicht zu klischeebehaftet ist, ob die Vorstellung des existenziellen Moments hinter dem Kofferpacken nicht ein stereotypes Bild der Flucht widerspiegelt. Vielleicht stellen wir, die wir nicht selbst geflüchtet sind, uns den Beginn der Flucht nur so vor? Vielleicht erschaffen wir diesen ikonischen Moment, um die Komplexität der Verhältnisse narrativ zu ordnen und in den Griff zu bekommen? Und welche anderen Situationen gehen dahinter verloren: Keine Zeit zu packen, Flucht nicht als einmaliges Ereignis, sondern als Hin und Her, kein starker persönlicher Bezug zu den Dingen etc.

Peter: Mir gab dieser Kommentar zu denken, ob wir mit unserem theoretischen Blick auf Dinge im Kontext von Flucht und Migration nicht auch in Fallen tappen – in Fallen des Reproduzierens von stereotypen Bildern. Und ob diese konzeptuellen Überlegungen vielleicht eher etwas mit uns und unseren Lebensbedingungen und Vorstellungen zu tun haben. Daher sind

greatest strength of our approach. This is a very lively and frequently touching method, which is nonetheless difficult to systematise. We do not say: Migration is like this or like this, but rather: Listen to the stories! That is the value of our research.

Friedemann: And this approach to research perhaps contains one possible answer to Andrea's question, which she posed regarding long-term solidarity: writing empathically and in a reflected way about migration, above and beyond large, attention-grabbing events. Equally, I consider the ability to reflect, in ethnography, on things, on personal stories, but also on the further-reaching structural, societal and material interrelations, to be foundational. When these elements are situated in good relationships with each other, then a (thing-oriented) ethnography can succeed, in my opinion.

Peter: In the Introduction to this volume, I began to think about the example of packing a suitcase, to illustrate the existential moment of selection. I was inspired by Jason de León's writings[7] on his encounter on the border between Mexico and the US, with migrants who had to take decisions regarding just such deliberations: What do I take with me? What do I need to survive, to get through Arizona's deserts that are hostile to life, to cross the border into the US, and to escape the clutches of the border police's subordinates? But also: Which things are worth remembering?

Joachim Baur: And these are undoubtedly decisive and make-or-break questions. But I also asked myself whether this moment of packing one's suitcase is not too burdened with clichés, and whether imagining an

wir ja auch dahin gekommen zu sagen, dass wir die Perspektive derer, die unsere Gesprächspartner:innen sind, in den Mittelpunkt stellen wollen, um auch erkennen zu können, dass unsere Projektionen und Konzepte korrekturbedürftig sind.

Andrea: Also mein Zugang, was die Existenzialität der Dinge angeht, ist ein biografischer. Meine Familie mütterlicherseits ist geflüchtet, und ich erinnere kein Familienfest, ohne dass Fluchtgeschichten erzählt wurden. Manche Geschichten haben mich zu verschiedenen Zeiten unterschiedlich angesprochen und berührt. Meine Oma ist mit vier Kindern geflohen, der Aufbruch war überstürzt und nur minimales Gepäck möglich. Sie selbst – so das Narrativ – hat nur das Familiensilber mitgenommen und Goldschmuck, weil sie wusste, sie muss später tauschen, gegen Brot und Butter bei Bauern unterwegs. Der älteste Sohn hat sich um die Papiere gekümmert und sich mit seinen achtzehn Jahren als so etwas wie der männliche Haushaltsvorstand verantwortlich gefühlt. Meine Mutter hat gar nichts mitgenommen und der Jüngste seinen Teddybären. Als Kind fand ich die Teddybär-Geschichte immer am ergreifendsten, und ich habe mir ausgemalt, wie auch ich in einem vergleichbaren Fall unbedingt meinen Teddy hätte mitnehmen wollen. Später fand ich die Familiensilber/-gold-Geschichte meiner Oma *tough*, funktionalistisch gut. Irgendwann gab es den Moment, als ich die rucksacklose Geschichte meiner Mutter bedenkenswert fand. Später hat sie Dinge gesammelt, die sie selbst gestickt, gehäkelt, genäht, gestrickt und gebastelt hat, und immer betont, dass sie das macht, damit ich sie später erben könne. Als

existential moment behind the physical act of packing might reflect a stereotypical image of flight. Perhaps we who have not had to flee ourselves, are just imagining the start of the process of flight to be like this – but in reality, it is not? Perhaps we are manufacturing this iconic moment to order the complexity of the relations involved into a narrative and to get a grip on them? And which other situations might be lost behind such a focus: no time to pack; flight not as a one-off event, but rather as a toing and froing; no strong personal connection to the things in question, etc.

Peter: Joachim's comment made me think whether our theoretical perspective on the things of flight and migration may make us walk into traps. The trap of reproducing stereotypical images, for example. And whether these conceptual deliberations might have more to do with us, with the conditions of our lives and with our notions and imaginings. We've thus reached the point of saying that we want to put the perspective of our interlocutors at the centre of our deliberations, also to recognise that our projections and concepts need correcting.

Andrea: My access to the existential significance that things have is a biographical one. My mother's side of the family had to flee, and I can't remember a single family gathering at which tales of flight were not told. Some stories spoke to me and moved me to varying degrees at various times. My grandma fled with four children, their departure hurried, and it was only possible for them to take a minimum amount of luggage. She herself – or at least so the narrative – took only the family silver and her

ich dann mein Elternhaus ausräumen musste, war ich überfordert von der Erblast der unzähligen Dinge. Ein paar Dinge habe ich eingepackt, wegen ihr. Aber ich verwende sie nicht, sie nehmen im Schrank nur Platz weg, eigentlich will ich sie gar nicht. Ich denke also ganz handfest und pragmatisch darüber nach: Was mache ich jetzt mit den Dingen meiner Mutter, die *ihr* so wichtig waren? Wie werde ich den Dingen ›gerecht‹, wenn ich mich von ihnen entlasten will?

Antonie: Da sieht man auch nochmal das, was Peter angesprochen hat, bei diesem Bild des kofferpackenden Menschen. Es geht ja ganz selten um einen einzelnen Menschen, sondern ganz häufig um Familien, um Menschen, die nicht allein, sondern mit ihren Nächsten fliehen. Wir reden also auch über Dinge, die uns mit anderen verbinden, wie jetzt in deinem Fall mit deiner Familiengeschichte. Aber ebenso im Moment des wie auch immer sich darstellenden Flüchtens oder Migrierens. Es wird ja nicht nur an die eigenen Dinge gedacht, die mitgenommen werden müssen, sondern häufig auch an die Dinge anderer: Was will mein Kind mitnehmen? Was will mein Partner oder meine Partnerin mitnehmen? Oder anders, was nehme ich jetzt mit von meiner Oma? Es sind ja ganz oft diese sozialen Kontexte, die immer wieder eine Rolle spielen. Eigentlich bei jeder Geschichte, die wir beschreiben, prägen zwischenmenschliche Beziehungen immer auch Mensch-Ding-Beziehungen.

Friedemann: Für mich ist in diesem Zusammenhang die Frage spannend: Was können Dinge? Was kann objektorientierte Ethnografie? Wo sind aber auch Grenzen und Schwierigkeiten? In der Forschung sind

gold jewellery, because she knew that she would have to barter later with farmers for bread and butter while en route. Her eldest son had taken care of the family papers and had felt, at age 18, to be something like the male head of the household. My mother took nothing with her, and her youngest sibling took only his teddy bear. As a child, it was this teddy bear story that gripped me most, and I painted a mental picture of how I would also have to take my teddy, at all costs, in a similar situation.

Later, I considered my gran's story about the family silver and gold to be *tough*, *good* from a functionalist perspective. There also came the point when I started reflecting on my mother's story of her fleeing "not carrying a rucksack", figuratively speaking. Later in life she collected things that she herself had embroidered, crocheted, sewn, knitted and handcrafted, things she always emphasized she had made so that I could inherit them later. When the time then came for me to clear out my parent's house, I was overwhelmed by the burden of inheritance the countless things possessed. I packed up a few things, because of her. But they take up space in my cupboards and I don't use them – I actually don't even want them. I try to think in a purely stalwart and pragmatic way about them: What should I do now with my mother's things, which were so important to her? How can I be "fair" to her things, if I want to unburden myself of them?

Antonie: Andrea's reflections again illustrate what Peter mentioned in this image of people packing suitcases. These images are rarely of a single person in isolation, but

gerade persönliche Dinge als Vermittler beispielsweise von Erinnerungen und Emotionen zweifellos häufig produktiv, wenn auch nicht immer unproblematisch. Gerade aber, wenn Dinge *für etwas* stehen sollen, besteht immer eine Gefahr von Verallgemeinerung, Klischees und ›Othering‹.[8] Will sagen: Dinge als Werkzeug der Erforschung von Mensch-Ding-Beziehungen sind das eine, Dinge als Stellvertreter solcher Beziehungen eine ganz andere Herausforderung. Die Frage lautet daher: Wo sind Dinge eher hinderlich? Wie müssen wir als Forschende Dinge einbetten oder auch über Dinge hinausgehen, um kritisch und reflexiv Zusammenhänge von Migration und Gesellschaft zu veranschaulichen? Sich dem zu stellen, empfand ich grade im Hinblick auf unsere Ausstellung[9] als eine Herausforderung. Der Koffer als ›Migrationsobjekt‹ ist dafür ein gutes Beispiel.

Joachim: Dem kann ich nur zustimmen. Als nahezu omnipräsentes, ikonisches, aber auch eingefrorenes ›Metasymbol‹ der Migration kann man dieses Ding, den Koffer, der in Ausstellungen zum Thema so oft gezeigt wird, ja wirklich nicht mehr sehen oder bringen. Die Frage ist: Gibt es andere Dinge, die hier freier sind? Oder sehen wir grundsätzlich ab von der Präsentation von Objekten im Kontext von Flucht und Migration, weil sie doch immer nur Klischees reproduzieren, zu kurz springen, sich ver- statt erschließen und damit komplexe Verhältnisse feststellen, fixieren? Oder aber finden wir andere Formen der Repräsentation, gar einen über die Repräsentation hinausgehenden Gebrauch des Ausstellungsraums, um Mensch-Ding-Beziehungen in ihrer Ambivalenz und Prozesshaftigkeit zu verhandeln? Das muss letztlich

much more often of families who do not flee alone but with their next-of-kin. Here we reach a point at which we're talking about things that connect us to others – in this case with your family history, Andrea. But which also connect us to the moment of flight or of migration, in whichever forms these actions manifest themselves. People do not merely remember their own things that they had to take with them, but also often things belonging to others: What does my child want to take? What does my partner want to take? Or the other way around: What do I now want to take with me from my gran? These social contexts that repeatedly play a major role incredibly often. In fact, in every story we describe, it is relations between people that always also shape relations between people and things.

Friedemann: In this regard, I'm excited about the following questions: What are things able to do? What is object-oriented ethnography able to do? Also where are its limits, and what difficulties does it face? In research, it's precisely personal things functioning as mediators, for example, as mediators between memories and emotions, that are undoubtedly often productive factors – even though they're far from being unproblematic. But when things are made to stand *for something*, the danger of generalisation, clichés and "othering" is always present.[8] What I want to say is: Things as a tool to research human-things relations are one challenge, but things as substitutes for human-things relations are quite a different challenge. This raises the question: When do things hinder research

konkret und praktisch ausprobiert werden, mit offenem Ausgang und ohne Garantie, dass es gelingt.

more than they help it? How, as researchers, do we have to implant things into contexts? But how do we also at times have to go beyond things, to visualize, critically and reflectively, connections between migration and society? I experienced confronting these questions as a challenge, particularly regarding our exhibition.[9] The suitcase as a "migration object" is a good example of this challenge.

Joachim: I can only second what Friedemann has said. People can simply no longer look at, or be presented with, this thing, the suitcase, which is so often shown in exhibitions on this subject, at least when this thing is reduced to an almost omnipresent, iconic, but also frozen "metasymbol" of migration. The question remains: Are there other things that are freer in this regard? Or should we turn away fundamentally from the presentation of objects in the context of flight and migration, because they only produce clichés after all, fall short, barricade rather than open things up, and with that attempt to secure and fix down complex relations. Or are we, alternatively, able to find other forms of representation or even use the exhibition space that goes beyond representation, to negotiate human-things relations in their ambiguity and in their processuality? Ultimately, this final option has to be experimented with, concretely and practically, in open-ended processes, and with no guarantee that it will work.

Anmerkungen

1. »MatMig« ist eine Abkürzung für den Projektnamen »Zur Materialität von Flucht und Migration«.
2. Mehr dazu kann in der folgenden Publikation nachgelesen werden: Bräunlein, Peter J.; Andrea Lauser: *Leben in Malula. Ein Beitrag zur Ethnographie der Alangan-Mangyan auf Mindoro (Philippinen)*, Pfaffenweiler 1993.
3. Zur kompletten Geschichte des Stück Stoffs: Brunner, Katharina; Samah Al-Jundi Pfaff (2020): »Das Stück Stoff«, abgerufen am 25. Oktober 2021, https://materialitaet-migration.de/objekte/the-piece-of-cloth/
4. Zum Inhalt dieser Zeitschriftenausgabe siehe: https://zeithistorische-forschungen.de/3-2018
5. Vgl. zum Beispiel das Interview mit dem »Kollapsologen« Pablo Servigne in *Philosophie Magazin*: Pablo Servigne im Interview mit Sven Ortoli: »Wir müssen Zusammenbrüche begleiten«, 18. November 2020, https://www.philomag.de/artikel/pablo-servigne-wir-muessen-die-zusammenbrueche-begleiten.
6. Zum Beispiel: Turnbull, Colin: *The Mountain People*, New York 1972 und Spittler, Gerd: *Handeln in einer Hungerkrise. Tuaregnomaden und die große Dürre von 1984*, Wiesbaden 1989.
7. Siehe dazu: De León, Jason: *The Land of Open Graves. Living and Dying on the Migrant Trail*. With photographs by Michael Wells, Oakland 2015.
8. Unter Othering wird das Andersmachen, Fremdmachen, Schaffen von Grenzziehungen zwischen den »Eigenen« und den »Anderen« verstanden. Siehe Quotrup Jensen, Sune 2011: Othering, Identity Formation and Agency. *Qualitative Studies*, 2(2), 63–78. https://doi.org/10.7146/qs.v2i2.5510
9. Mehr zur Ausstellung unter: http://www.forum-wissen.de/

Notes

1. "MatMig" is the abbreviation for the project "On the Materiality of (Forced) Migration".
2. More on this issue in: Peter J. Bräunlein; Andrea Lauser: *Leben in Malula. Ein Beitrag zur Ethnographie der Alangan-Mangyan auf Mindoro (Philippinen)*, Pfaffenweiler 1993.
3. For the complete story of the piece of cloth, see Katharina Brunner and Samah Al-Jundi Pfaff: "The Piece of Cloth", accessed 25 October 2021, https://materialitaet-migration.de/en/objekte/the-piece-of-cloth/
4. For the contents of this issue of the magazine, see https://zeithistorische-forschungen.de/3-2018
5. Cf. for example the interview with the "collapsologist" Pablo Servigne in *Philosophie Magazin*, conducted by Sven Ortoli: "Wir müssen Zusammenbrüche begleiten", 18 November 2020, https://www.philomag.de/artikel/pablo-servigne-wir-muessen-die-zusammenbrueche-begleiten
6. For example, Colin Turnbull: *The Mountain People*, New York 1972, and Gerd Spittler: *Handeln in einer Hungerkrise: Tuaregnomaden und die große Dürre von 1984*, Wiesbaden 1989.
7. Jason De León: *The Land of Open Graves: Living and Dying on the Migrant Trail*. With photographs by Michael Wells, Oakland 2015.
8. "Othering" is understood to mean making other things or people appear to be different or foreign, and the drawing of boundaries between one's "own" people and "other" people. See Sune Quotrup Jensen: Othering, Identity Formation and Agency. *Qualitative Studies*, 2(2), 2011, pp. 63–78. https://doi.org/10.7146/qs.v2i2.5510.
9. More on the exhibition at: http://www.forum-wissen.de/

Foodways

Friedemann Yi-Neumann

Foodways

Abb. 1: Injera, 2021. ©Hannah Bohr

Image 1: Injera, 2021. ©Hannah Bohr

Wenn Menschen migrieren, migrieren Dinge mit ihnen. Ein wichtiger Aspekt dieser mobilen materiellen Kulturen sind Nahrungsmittel, die im oft schwierigen Migrationsalltag neue Relevanz entfalten. Die Mobilität des Essens stellt beileibe kein neues Phänomen dar. Spätestens mit dem Zeitalter des Imperialismus und Kolonialismus veränderte der Welthandel Gesellschaften auf der ganzen Welt in Bezug auf das Essen, wie etwa Sidney W. Mintz (1985) in seinem Buch *Sweetness*

When people migrate, things migrate with them. Foodstuffs are an important aspect of mobile, material cultures and often take on a new relevance in the obstacle-strewn everyday life of migrants. That said, the mobility of food is in no way a new phenomenon. As Sidney Mintz demonstrated in his book *Sweetness and Power – The Place of Sugar in Modern History* (1985) through the case study of sugar as a colonial good, world trade changed relations

and Power – The Place of Sugar in Modern History am Beispiel der Kolonialware Zucker dargelegt hat. Die Speisekarten urbaner Metropolen wären ohne historische Entwicklungen wie diese kaum denkbar. Die meisten Nahrungsmittel, die wir im Westen täglich zu uns nehmen, sind substanziell mit Migration verflochten, sei es aufgrund von Handelswegen oder der Lebensmittelproduktion durch Migrant:innen. In diesen kulinarischen Verflechtungen finden sich vertraute und unbekannte Geschmäcker, jedoch auch altbekannte Ungleichheiten wieder.[1]

In diesem Kapitel geht es am Beispiel des *Injera*-Brots sowohl um die Relevanz von Essen für Menschen im Kontext von Flucht und Migration als auch um Nahrungsmittel, die migrieren, wie anhand des Teff-Mehls gezeigt wird, aus dem dieses Brot besteht. Im Zuge von Mobilität und Migration verändern sich die Zutaten, die Zubereitung und der Geschmack von Speisen. Damit einhergehend verändert sich auch der soziale und gesellschaftliche Stellenwert des Essens, wie am Beispiel »Curry« zu sehen ist, sowie der Personen, die es verzehren, was im Abschnitt »Chapati« behandelt wird. Diese Neupositionierungen sind dabei keineswegs unumstritten. Gerade auch im Kontext von Mahlzeiten treffen manchmal unterschiedliche Vorstellungen, Bedingungen und Bedürfnisse aufeinander, was am Beispiel »Geschirr und Fremdbestimmung« deutlich wird.

to food in societies across the globe. Further, processes of imperialism and colonialism accentuated these changes. It is hard to imagine how the menus of modern metropolises would read today if these developments had not taken place. Most foodstuffs we consume regularly in the West are substantially interwoven with migration processes, whether in the form of trade routes or of food production by migrants. These culinary overlappings allow familiar and previously unknown tastes to be detected and often-explored injustices to be discerned.[1]

This chapter takes the case of *injera* bread to examine the relevance of food to people in the context of flight and migration, and to study foodstuffs that migrate themselves – like the *teff* flour from which this bread is made. The ingredients, the preparation methods and the tastes of dishes change as a result of mobility and migration. Alongside this, the social and societal values attached to particular dishes or foods and the people who dine on them also changes, as can be seen in the example of "curry", which is dealt with in the section titled "chapati". These new positionings of food in relation to value are certainly contested. In the context of mealtimes, conflicting notions, conditions and needs collide with each other. These contradictions are addressed in the section below about conflicts and heteronomy connected with tableware.

Injera

Dass bestimmte Nahrungsmittel im Zuge von Migration für Menschen wichtig werden können, obwohl diese zuvor keinen besonderen Bezug zu diesen Nahrungsmitteln oder deren Zubereitung hatten, wird hier am Beispiel Desalegns und seinem Bezug zu *Injera* und äthiopischem Essen verdeutlicht. Desalegn kommt aus Addis Abeba, Äthiopien, und ist 2018 nach Europa geflüchtet. Während seiner Zeit in einer Frankfurter Geflüchtetenunterkunft – er befand sich als »Dublin-Fall« in einem wenig aussichtsreichen Asylverfahren – hatte er einmal mit einem Zimmergenossen etwas Geld gespart, um der Tristesse des Alltagslebens in der Unterkunft und dem schlechten Essen zu entfliehen. Beide waren in ein äthiopisches Restaurant gegangen, weil sie gerne einmal wieder »wie zu Hause« essen wollten. Doch sie waren überrascht über das, was ihnen dort als *Injera* serviert wurde. Es schmeckte völlig anders, als sie es kannten, und bestand, wenn überhaupt, nur aus sehr wenig *Teff* (einer Hirseart, siehe unten). Das Essen war teuer, und abgesehen von der Brotfrage, »einigermaßen okay«, wie Desalegn es ausdrückt. Doch die Mahlzeit überzeugte beide nicht besonders; es war schlicht nicht das, was sie erwartet hatten. Satt, aber etwas desillusioniert, verließen sie das Restaurant.

Für Desalegn war nach seinem Besuch im Restaurant klar, dass er, wenn er »richtiges« Essen haben möchte, es selbst kochen müsste. Erstens, damit es richtig schmeckt und zweitens, weil es für ihn, wie für die Mehrzahl der Menschen im Asylverfahren, unerschwinglich ist, regelmäßig ins Restaurant zu gehen. Für seine Freund:innen kocht

Injera

The issue of particular foods becoming important during the course of migration – even when no strong connection to the same foods had previously existed – is concretised here through the case study of Desalegn and his relationship with Ethiopian food. Desalegn comes from Addis Ababa in Ethiopia, and fled to Europe in 2018. While staying in a Frankfurt refugee accommodation, after having applied for asylum in a case that the Dublin Regulation made look far from promising, he and a roommate once saved up money to get away from the dreariness of everyday life in the refugee housing – and the bad food that accompanied it. Both went together to an Ethiopian restaurant, because they wanted to eat food again "like the food at home". Yet, they were surprised at what was served up to them under the name of *injera*. It tasted utterly different from how they knew it, and if it contained any *teff* – a variety of millet – at all, then only very little. The food was expensive and, apart from the bread question, "more or less OK", as Desalegn put it. But neither of them was really convinced by the meal's authenticity – it simply wasn't what they had expected. Satiated but somewhat disillusioned, they left the restaurant.

After going to the restaurant, it became clear to Desalegn that if he wanted to have the "right" kind of food, then he would have to cook it himself. First, so that it would taste right, and second, because he – like most people seeking asylum – simply couldn't afford to go regularly to restaurants. From then on he regularly started

Foodways

er nun regelmäßig *Injera*-Gerichte, die bei seinen Gästen sehr gut ankommen. Essen ist für Desalegn im Exil wichtig geworden, obwohl er keineswegs eine ausgeprägte Beziehung zum Kochen hatte.

Kochen im Exil ist ein Prozess der experimentellen Neuaushandlung und Neupositionierung, in dem Menschen sich mit ihren Fähigkeiten einbringen, neue Netzwerke schaffen und ein gewisses Maß an Selbstbestimmung erreichen können (Verdasco 2022). Lebensmittel ermöglichen dabei nicht nur das bloße physische Überleben, wie die Bezeichnung es suggeriert. Sie sind auch ein Mittel, um (geschmackliche) Bezüge zur Vergangenheit und Zugehörigkeiten in der Gegenwart herzustellen und neue Verbindungen zu knüpfen. Essen kann, obwohl es keineswegs statisch ist, geschmackliche Kontinuität über Grenzen und Zeiträume hinweg ermöglichen. Der vertraute Geruch von Essen aus der Kindheit oder die Netzwerke rund um das Essen von Geflüchteten mögen hier als Beispiel dienen (Povrzanović Frykman 2018).

Doch Kochen ist oft kein grenzenloses Experimentieren auf neutralem Terrain. Schwierigkeiten verursacht nicht nur das Fehlen von Geld, Ressourcen, Räumen und Equipment, sondern auch mangelnde Wertschätzung und sogar Abwertungen. Gesellschaftlich dominante Wahrnehmungsweisen und Deutungshoheiten sorgen etwa dafür, dass einige Speisen und Gerüche als »konventionell«, andere als »exotisch« gelten. Wieder andere werden ganz abgelehnt. Diese Ablehnung kann rassistisch begründet sein und betrifft vor allem auch die Menschen, die mit diesen Gerichten und Gerü-

cooking *injera* dishes for his friends, which were very much appreciated. Food has become important for Desalegn in exile, even though he previously did not have a deep relationship to cooking.

Cooking in exile is a process of experimental new negotiation and new positioning, in which people can make use of the abilities they have, create new networks and can achieve a certain degree of autonomy (Verdasco, 2022). In this process, foodstuffs do not merely serve the need of simply surviving, as their name might suggest; rather, they are a means of establishing connections of taste to the past, relations of belonging in the present and creating new connections. While food is in no way a static quantity, it can facilitate continuities of taste across borders and time. The familiar smell of food from childhood, or the networks connected to the refugees' food, can serve here as examples (Povrzanović Frykman, 2018).

But cooking is often not a borderless experiment, conducted on neutral terrain. It is not merely the absence of money, resources, equipment and places to cook and eat in that presents difficulties, but also that this type of cooking has a generally low status or even put down actively by negative attitudes towards it. Ways of perceiving, and interpretational sovereignties that are dominant in certain societies, determine that particular dishes and smells are considered "conventional", while others are experienced as "exotic". Some dishes and smells are even rejected entirely. Such rejection sometimes has racist roots and affects primarily those who are associated with these dishes and

chen assoziiert werden, wie Ethnolog:innen dies etwa am Beispiel von *Berbere*, einer vielverwendeten Gewürzmischung der äthiopischen und eritreischen Küche herausgearbeitet haben.[2] Ein ethnographischer Blick auf die materielle Kultur von Koch- und Essenspraktiken kann die soziale Positionierung von Menschen und Dingen sowie Prozesse der Veränderung nachvollziehbar machen.

Injera ist ein Fladenbrot, das gemeinhin als Grundlage der äthiopischen Küche gilt. Das Sauerteigbrot besteht meist aus *Teff*, einer Hirseart, die oft mit jeweils anderen (lokalen) Getreideformen gemischt wird. Desalegn ist in Frankfurt nicht nur mit dem Brot der Restaurants selten zufrieden, sondern auch mit dem *Injera* in einschlägigen Läden der eritreischen und äthiopischen Communitys der Stadt, die er regelmäßig besucht. Ein Bekannter hat ihm jedoch eine junge *Injera*-Bäckerin empfohlen. Diese stellt *Injera* nach seinem Geschmack bei sich zu Hause her. Der Teig ist keineswegs schnell gemacht, denn er muss zwei bis drei Tage gehen, bis er seinen mildsäuerlichen Geschmack entwickelt. Danach werden die Fladen entweder in einer Pfanne oder mit Hilfe eines speziellen Geräts, das einem Crêpes-Eisen ähnelt, zubereitet. Da sich auch seine Freund:innen in Frankfurt einig sind, dass diese Frau das beste *Injera* in Frankfurt macht, bezieht er es nur noch von ihr.

Teff

Doch was ist eigentlich *Teff*? *Teff* ist eine Pflanze aus der Familie der Süßgräser, auch Zwerghirse genannt. Dabei handelt es sich um eine

smells. Social anthropologists have also explored such racist connections by looking at diverging reactions to *berbere*, a spice-mixture often used in Ethiopian and Eritrean cuisine.[2] An anthropological perspective on the material culture of cooking and eating practices can make the social positioning of people in relation to food comprehensible, just as it can illuminate processes of change regarding this positioning.

The flatbread *injera* is indisputably one of the stables of Ethiopian cookery. The sour-dough bread mostly consists of *teff*, a kind of millet, often with flours from other, local forms of grain mixed in. In Frankfurt, Desalegn was rarely satisfied with either the bread in restaurants or the *injera* sold in the leading shops for the Eritrean and Ethiopian communities in the city, which he visited regularly. At some point, however, an acquaintance recommended a young, female *injera* baker to Desalegn. Created at home in the baker's kitchen, this *injera* is to Desalegn's taste. The dough rises in a two-to-three-day process, during which it develops its mild, sour taste. Once the dough has risen, the flatbreads are made either in a frying-pan or with the help of a special kitchen device, similar to a crêpes-maker. With all his friends agreed that this woman makes the best *injera* in Frankfurt, Desalegn now buys *injera* only from her.

Teff

But what is *teff*, actually? *Teff* belongs to the *Poaceae* family of grasses. It is a type of millet, a basic foodstuff in large parts of

Foodways

Abb. 2: Teffmehl, 2021. ©Hannah Bohr

Image 2: Teff, 2021. ©Hannah Bohr

Hirsesorte, die in weiten Teilen Äthiopiens und Eritreas ein Grundnahrungsmittel ist. Das bedeutet jedoch nicht, dass diese Hirse dort günstig ist; sie stellt vor allem auf dem Land eher ein Luxusgut dar.

Zu Hause im urbanen Zentraläthiopien hatte Desalegn nur selten gekocht. Während der langen Arbeitstage in seinem Job im Verkehrswesen und nach Feierabend aß er *Teff-Injera* meist in einem Straßenrestaurant mit seinen Freund:innen. Manchmal besuchten sie sich auch gegenseitig und aßen gemeinsam bei einem:r von ihnen zu Hause, wo meist die Frauen das Essen zubereiteten. Nur selten kochte er selbst für andere und dann meist einfache Gerichte.

Lange weitgehend unbekannt, hat *Teff* mittlerweile auch Bedeutung in der westlichen Welt erlangt, jedoch auf eine andere Art und Weise. Hier wird es als vielseitiges, glutenfreies und mineralstoffreiches »Superfood« gefei-

Ethiopia and Eritrea. This does not mean, however, that this form of millet is cheap. On the contrary, it has rather the status of a luxury product, particularly in the countryside.

Back home in urban, central Ethiopia, Desalegn rarely used to cook. During the long working days in his job in the transportation industry, and after finishing work in the evenings, he used to eat *teff-injera* in a fast-food restaurant with his friends. They also occasionally visited each other's homes, where the women usually prepared the food. He rarely cooked for himself or others, and when he did, mostly simple dishes.

Long relatively unknown outside Eastern Africa, *teff* has now become a significant foodstuff in the Global North, for remarkable reasons. In the West *teff* is celebrated as a multifacetted, gluten-free, mineral-rich "superfood" (Cheng et al., 2017).

ert (Cheng et al. 2017). *Teff* ist zum begehrten Ernährungsbestandteil einer gesundheitsbewussten Mittelschicht geworden. In Frankfurt vertreiben manche Biobäcker *Teff*-Brötchen als Teil ihres Sortiments. Seine Eigenschaften ermöglichen dem Getreide eine beeindruckende Karriere und einen ebenso beeindruckenden Preis. Ein Kilo *Teff*-Mehl kostet im Onlinehandel in Europa zwischen 13 und 15 Euro (gegenüber ein bis zwei Euro pro Kilo Weizenmehl).

Lebensmittel sind mobil. Sie werden weltweit verkauft, transportiert und verwendet. Während *Teff* im Globalen Norden als »hipp« und »gesund« gilt, lassen sich anhand der Mobilität dieses Getreides auch die ungleiche Verteilung von Ressourcen und die globalen, teils neokolonialen Machtverhältnisse aufzeigen, die unter anderem ein Grund dafür sind, warum Menschen migrieren.

»Teffgate«: Der große Betrug am äthiopischen Staat

Das Mehl des Süßgrases ist keineswegs nur so teuer, weil es sich um ein hochwertiges Nahrungsmittel handelt. Es gibt dafür auch patentrechtliche und handelspolitische Gründe. Eine niederländische Firma hat sich 2002 ein Patent zur *Teff*-Verarbeitung sichern lassen. Ferner schloss sie einen Vertrag mit der staatlichen *Ethiopian Agricultural Research Organization* (EARO) ab, die dem Unternehmen freien Zugang zu diesem äthiopischen Getreide ohne finanzielle Gegenleistung gewährte. Die Firma gehört zu einem undurchsichtigen Konglomerat an Unternehmen, die im Besitz des Niederländers Jans Roosjen sind.[3]

Diese Vereinbarung wurde 2019 von

Teff has become a desirable nutritional component for a health-conscious middle class. In Frankfurt, some organic bakers sell *teff*-rolls as part of their sales mix. These characteristics have facilitated an impressive career for the grain – with impressive prices to match. One kilo of *teff* flour, traded online in Europe, costs between 13 and 15 euros – compared to just one or two euros for a kilo of wheat flour.

Foodstuffs are mobile. They are sold, transported, used and consumed worldwide. Considered "trendy" and "healthy" in the Global North, the mobility of this grain enables a spotlight to be thrown on both the unjust distribution of resources, and on the global and partly neocolonial power relations, which are one of the reasons why people migrate.

"Teffgate": The Great Defrauding of the Ethiopian State

Flour made from this *Poaceae* grass is not expensive merely because it is highly nutritional. Rather, patent law and trade politics were decisive in pushing up *teff* prices. In 2002, a Dutch company took out a patent for the processing of *teff*. Moreover, it also concluded a contract with the Ethiopian Agricultural Research Organization (EARO), which granted the company free access to this Ethiopian grain without having to pay for this substantial economic advantage. The company is part of an opaque conglomerate of enterprises, all belonging to the Dutch owner Jans Roosjen.[3]

This agreement was analysed by Zech-

Zecharias Zelalem in einem umfangreichen Beitrag für den *Addis Standard* unter dem Stichwort »Teffgate« aufgearbeitet und darin als Betrug am äthiopischen Staat bezeichnet (Zelalem 2019). Es wurde darin nicht nur bemängelt, dass Roosjen freier Zugang zum Saatgut gewährt würde, sondern vor allem, dass er sich mit Hilfe des Patents als »Erfinder der *Teff*-Verarbeitung« präsentierte und jegliche weitere Patente auf Grundlage des Getreides ausgeschlossen habe.[4] Dies verursachte eine große Empörung weit über Äthiopien hinaus. Doch auch auf der äthiopischen Seite wurde Dr. Demel Teketay, damaliger Chef der EARO, wegen seiner Zustimmung zu besagtem Deal kritisiert. Denn neben der kulturellen Aneignung der *Teff*-Produktion monopolisierte das Patent den internationalen *Teff*-Handel und machte es Äthiopien unmöglich, eigene Handelsverträge diesbezüglich abzuschließen und an den Gewinnen der *Teff*-Produkte beteiligt zu sein.

Ein niederländisches Gericht kippte Roosjens Patent im November 2018 aufgrund »mangelnder Innovationskraft«; die vertraglich festgehaltenen Rechte an der *Teff*-Verarbeitung behielt die Firma jedoch weitgehend. Das heißt, dass sich an den Verhältnissen ebenso wie an den Preisen in absehbarer Zeit wohl wenig ändern wird und Äthiopier:innen auch weiter eine angemessene finanzielle Beteiligung an den durch *Teff* erwirtschafteten Gewinnen verwehrt bleibt. Migration ist komplex und hat vielfältige Gründe – die Folgen ökonomischer Ausplünderung durch Unternehmen für Menschen in Äthiopien gehörte sicherlich dazu.

arias Zelalem in a comprehensive article for the *Addis Standard* in 2019, which characterized the deal as "Teffgate" and judged the contract to be a fraud against the Ethiopian state (Zelalem, 2019). Alongside the substantial argument that Roosjen had been given free access to Ethiopia's seed diversity, the greatest charge against him was that he had used the patent to present himself as the "inventor of teff processing" and excluded any further patents based on the grain.[4] The outrage Roosjen's strategy unleashed reverberated far beyond Ethiopia itself. Meanwhile, in the concrete Ethiopian context, Dr Demel Teketay, former head of the EARO, was criticised for agreeing to the aforementioned deal. Beyond the cultural appropriation of *teff* production, the patent also monopolised the international *teff* trade, making it impossible for the Ethiopian state to negotiate its own trade agreements for the grain and not granting Ethiopia a share in the profits from *teff* products.

A Dutch court overthrew Roosjen's patent in in November 2018, judging that it contained a "lack of innovative power"; nonetheless, the company largely retained the contractually agreed rights to *teff* processing. Meaning that, with regard to the economic, social and price relations of *teff*, little will change in the foreseeable future: Ethiopians will continue to be denied an appropriate share in *teff* profits. Among the complex and manifold reasons for emigration out of Ethiopia, it would be ludicrous to deny that the human consequences of economic exploitation are significant.

Abb. 3: Zutaten, 2021. ©Hannah Bohr, 2021   Image 3: Ingredients, 2021. ©Hannah Bohr

Kochen im Exil/Cooking in Exile

Globale Bedingungen des Handels, Flucht und Migration kommen beim Essen im Exil zusammen. Für das Essen, das auf den Bildern in diesem Beitrag zu sehen ist, haben Desalegn und seine Freund:innen zusammengelegt. Seit er nicht mehr in der Unterkunft lebt, ist er in einem Umfeld, das seine Gerichte schätzt, statt sich über ungewohnte Gerüche zu beschweren. Da in seinem näheren Freund:innenkreis jedoch niemand äthiopisch kochen kann, musste er sich dieses Wissen selbst aneignen. Mit Hilfe von *Youtube*-Tutorials wagte sich Desalegn in Frankfurt als Hobbykoch an die durchaus komplexe Küche.

Doch Desalegn betont, dass er es niemals mit erfahrenen äthiopischen Köch:innen aufnehmen könne. Denn manche Gerichte benö-

Cooking in Exile

The global conditions of trade, flight and migration all intersect in the materiality of food and exile. Desalegn and his friends all contributed to providing the food and dishes that were photographed for this article. After leaving the refugee accommodation, Desalegn began living in a milieu that valued his dishes, instead of complaining about their unusual aromas. But because no one in his circle of closer friends is able to cook Ethiopian food, he had to gather this knowledge himself. Assisted by YouTube tutorials, Desalegn dared, in Frankfurt, to train as an amateur cook in this unquestionably intricate cuisine.

Yet Desalegn emphasises that he could never compete with experienced Ethiopian

tigen Tage der Vorbereitung und ihre Zubereitung beinhaltet zahlreiche Arbeitsabläufe, wie etwa *Dor Wot*, ein sehr aufwändiger pikanter Hähncheneintopf, der in Kochbüchern nur in vereinfachter Form für ein westliches Publikum zu finden ist. Im Rhein-Main-Gebiet leben einige Expert:innen der äthiopischen Küche. Von Zeit zu Zeit fragt Desalegn sie telefonisch um Rat. Sich an den Rezepten von Zuhause zu versuchen, ermöglicht Ablenkung, Genuss und neues Selbstbewusstsein in einem Alltag, der für Geduldete wie Desalegn ansonsten von großer Unsicherheit und Perspektivlosigkeit geprägt ist.

Zu der Grundlage *Injera* kocht Desalegn gewöhnlicherweise *Mesir Wot* (ein scharfes Linsengericht), *Sh'ro Wot* (Sauce aus Sauerbohnen) und dazu *Alicha Wot* (Kartoffel- und Karotteneintopf), Spinat, Bohnen,

cooks. Indeed, some dishes require days of preparation and numerous work-processes, like *Dor Wot*, for example, an extremely elaborate and spicy chicken casserole is only found in cookbooks in a form simplified for Western readership. A number of experts for Ethiopian cuisine live in the Rhine-Main region, and, now and again, Desalegn phones them up while he's cooking, to ask for their advice. Attempting recipes from home is a welcome distraction and a source of pleasure and new self-confidence for those, like Desalegen, classified as merely "tolerated" [*geduldet*][5] by the German asylum process: effectively a suspension of deportation. This status is understandably characterised by extreme uncertainty and the absence of prospects.

To accompany *injera* as a staple, De-

Abb. 4: Berebere (rot), äthiopischer Pfeffer (braun) Shiro-Mehl (gelb), 2021. ©Hannah Bohr

Image 4: Berebere (red), Ethiopian pepper (brown), and Shiro flour (yellow), 2020. ©by Hannah Bohr

Kohlsalat, manchmal Fleisch. Eine wichtige Zutat ist *Berebere*, eine rote und pikante Gewürzmischung. Diese wird aus Chillischoten, Koriander, Knoblauch, Basilikum und anderen Gewürzen hergestellt und gilt mit ihrer würzigen Schärfe gemeinhin als zweiter Grundbestandteil der äthiopischen und eritreischen Küche.

Umgangssprachlich wird Essen, das aus verschiedenen Gerichten zusammengestellt ist, *Maebrawi* genannt. Das bedeutet auf Amharisch, der am zweithäufigsten gesprochenen Sprache (und Amtssprache) Äthiopiens, eigentlich »sozial«. So wie die unterschiedlichen zusammengestellten Gerichte das Mahl ergeben, bringt das Essen die Menschen zu einem gemeinsamen Mahl zusammen. Doch gerade dieses Zusammenkommen ist im Exil eine Herausforderung und in der Zeit

salegn usually cooks *Mesir Wot* (a spicy lentil dish), *Sh'ro Wot* (a sauce made from pickled "French" beans – *Phaseolus vulgaris*) – served alongside *Alicha Wot* (a potato and carrot stew), spinach, beans, coleslaw and sometimes meat. *Berebere* is an important ingredient, an orange-red, hot-spice mixture made from chilli peppers, coriander, garlic, basil and other spices. It is considered, with its tangy spiciness, to be the second core substance in Ethiopian and Eritrean cookery.

In colloquial usage, a meal that's put together from different dishes is called *maebrawi*, which means "social" in Amharic, the second most widely spoken language in Ethiopia – and its state language. Just as a meal can be created from dishes laid out alongside each other, food brings

Abb. 5: Essenszubereitung, 2021. ©Hannah Bohr

Image 5: Preparing food, 2021. ©Hannah Bohr

Foodways

Abb. 6: Injeraplatte, 2021. ©Hannah Bohr

Image 6: Injera smorgasbord, 2021. ©Hannah Bohr

der COVID-19-Pandemie eine Praxis, die nur schwer umzusetzen ist.

Weil *Injera*-Gerichte für Desalegn in Deutschland aber doch eher die Ausnahme sind, da sie viel Arbeit verursachen, die Zutaten nicht billig und nicht immer einfach zu bekommen sind, hat er angefangen zu experimentieren. Um regelmäßig einen Geschmack von ›Zuhause‹ zu schmecken, versucht er, in Deutschland gängige Gerichte auf seine Art zu adaptieren. *Berebere* eignet sich etwa sehr gut als Gewürz in Tomatensaucen, sei es für Pasta oder Pizza, und gibt diesen Gerichten eine pikante Note. Zudem verwendet er es vermischt mit Olivenöl, um Toast, aber auch Risotto, ein bisschen würziger und geschmacklich vertrauter zu machen.

Nachdem er regelmäßig Erfahrungen mit dem Kochen äthiopischer Speisen gesammelt

people together to share that meal. But it's precisely this coming together that is a major challenge when living in exile – and one that's hard to realise during the Covid-19 pandemic.

But because *injera* dishes remain the exception rather than the rule for Desalegn in Germany, and because they involve a lot of work, with ingredients that aren't cheap and that are often difficult to obtain, he has begun to experiment. To regularly get a taste of 'home', he tries to adapt dishes that are often eaten in Germany to his own cuisine. *Berbere* turns out to be a very welcome addition as a spice for tomato sauces, whether for pizza or pasta, lending these dishes a sharper note. Desalegn also uses *berbere* mixed with olive oil, on toast or for risotto, to make a whole meal a bit

hat, hat sich Desalegn vorgenommen, auch einmal selbst *Injera* zu machen.

Während die Geschichte von Desalegn zeigt, wie Nahrungsmittel für Menschen im Exil einen neuen Stellenwert einnehmen und welche globalen Verbindungen von Migration und Essensökonomien bestehen, rückt im Folgenden die Migration von Nahrungsmitteln ins Zentrum. Gerichte legen komplexe, oft unbekannte Wege zurück und werden in Form und Geschmack transformiert. Damit verändert sich auch, wo und von wem sie begehrt und wie sie verzehrt werden. Geschätztes und vermisstes Essen muss dabei nicht unbedingt das sein, das qualitativ besonders hochwertig ist.

Curry
von Eungso Yi

Als Kind wusste ich gar nicht, dass Curry ein indisches Essen ist. Curry war bei uns eins der vielen Gerichte, für das ein bestimmtes Gewürz mit Gemüse gekocht und mit Reis gegessen wurde. In der Schule habe ich gehört, dass Curry aus Japan kommt. Später habe ich doch erfahren, dass Curry aus Indien kommt, dennoch kannte ich nur koreanisches Curry. Heute sind viele internationale Gerichte in Südkorea verbreitet und sehr beliebt, aber damals gab es diese nur in wenigen Restaurants. In dem Viertel Seouls, wo sich die Fremdsprachenuniversität befindet, an der ich damals (2005) studierte, gab es jedoch einige Restaurants mit authentischen internationalen Gerichten. Als Germanistikstudentin wurde ich manchmal dazu gedrängt, meinem Fachbereich entsprechend,

spicier and to give it a taste he's familiar with.

And now that he's built up experience in cooking Ethiopian dishes, Desalegn has resolved to bake *injera* himself, at least once.

Desalegn's story demonstrates how foodstuffs acquire a new value for people in exile, but it also illustrates which global connections exist between migration and food economies. The following chapter highlights how foodstuffs migrate themselves. Dishes often leave a trail of complex and often unknown routes behind them and are themselves transformed in their form and their tastes through processes of migration. These migrations also alter where and who desires which dishes as well as how various dishes are actually eaten. The kind of food that individuals and groups value and miss is not necessarily that consisting of what, seen more objectively, could be categorised as especially high-quality food.

Curry
by Eungso Yi

As a child, I didn't know that curry is an Indian dish. For us, 'curry' was just one among several dishes, for which vegetables were cooked with a particular spice and eaten with rice. At school, I heard that curry came from Japan. Later I discovered that curry comes from India, but I only knew the Korean variety. These days lots of international dishes are widespread and extremely popular in South Korea, but at that time they could only be found in a hand-

im Restaurant ›Bismarck‹ Schnitzel und Würste zu verzehren. Nach einiger Zeit, in der ich nur koreanisches, japanisches und thailändisches Curry gegessen habe, habe ich es doch irgendwann geschafft, zum ersten Mal die indische Version zu essen. Erst dann wurde mir klar, dass es so viele Varianten von Curry gibt und dass Curry in den indischen Sprachen einfach ›Gewürze‹ bedeutet.

Seitdem ich in Frankfurt wohne, kann ich jederzeit vielfältige Currygerichte genießen. Wenn die Zeit kommt, in der ich langsam Heimweh bekomme, koche ich in meiner Wohngemeinschaft ein Curry. Kommt jemand dazu und fragt mich, was ich koche, so denke ich mir:

> Als die Brit:innen Indien kolonisierten, du weißt das ja bestimmt, kam das Curry aus Indien nach Großbritannien. Als dann in Japan die Meiji-Restauration stattfand und Japan die Kultur der westlichen Länder umfangreich aufnahm, kam das Curry von Großbritannien nach Japan. In Japan wurde die Aussprache von Curry in ›Ka-re‹ geändert. Japan entwickelte eine eigene Curryform, und die wurde wiederum von den Japaner:innen nach Korea gebracht, als Japan Korea kolonisierte. Nach der Befreiung von der japanischen Besatzung wurde es in Korea als Instant-Lebensmittel kommerzialisiert und in Haushalten und Kantinen von Schulen, Krankenhäusern und Militäreinrichtungen populär. Was du heute Abend essen wirst, ist diese koreanische Curryvariante, die von Indien durch Großbritannien und Japan in Südkorea gelandet ist, bei der der authentische Geschmack durch Stärke verdünnt wird.

ful of restaurants. In the district of Seoul that houses the Foreign Studies University, where I was studying at that time, in 2005, there were such restaurants with authentic international dishes. As a German literature student, I was pressured, because of my choice of degree courses, into consuming *schnitzel* and *würste* [sausages] in the 'Bismarck' restaurant on several occasions. After a long phase in which I only ate Korean, Japanese and Thai curry, I eventually managed to eat the Indian version for the first time. It was only then that I started to understand that there are countless varieties of curry, a word that in Indian languages simply means 'spices'.

Since I moved to Frankfurt, I've been able to enjoy a multitude of curry dishes at almost any time of the day or night. On those days when I feel homesick do I cook a curry in my shared flat. If someone were to show up here and ask me what I'm cooking, this is what I would *think*:

> When the British began to colonise India, curry moved from India to Great Britain – as you likely know already. And in that period, in which the Meji Restoration took place in Japan, the late 1860s, and Japan gained extensive access to the culture of Western countries, curry then moved from Great Britain to Japan. On arrival there, its pronunciation changed from curry to *'Ka-re'*. Japan developed its own form of the dish, and Japanese people went onto import it into Korea, when they colonised that country. After the liberation from the Japanese occupa-

Weil diese Antwort zu kompliziert und langwierig ist, entgegne ich meistens nur kurz: »Curry, nach koreanischer Art.«

Chapati

Ausgehend von der sich wandelnden Bedeutung und der globalen Zusammensetzung von Mahlzeiten wird im folgenden Abschnitt das Essen als Mittel der sozialen Interaktion behandelt. Der griechische Philosoph Sokrates (470–399 v. Chr.) schätzte »gute Mahlzeiten« und wusste bereits um ihre Fähigkeit, »gute Menschen« zusammenzubringen. Doch wie verhält es sich mit gemeinsamen Mahlzeiten in Unterkünften für Geflüchtete, Einrichtungen, die von Fremdbestimmung geprägt sind und in denen Menschen unterschiedlicher Herkunft unfreiwillig in erzwungener Nähe zusammenleben? Wie gemeinsame Mahlzeiten Menschen in Aufnahmezentren zusammenbringen können, ist Thema des nächsten Abschnitts.

Wir befinden uns im Durchgangslager Friedland im Winter 2018/2019. Für gewöhnlich bereitete Arif Ibrahim *Dal* (Linsensuppe) oder Curries aus Kartoffeln und Gemüse zu. Sein Freund Raafe war für die Zubereitung der *Chapatis* (ungesäuertes flaches Brot) zuständig, und Haani bereitete den Raum für das gemeinsame Essen vor und machte danach den Abwasch. Arif briet das Hühnchen mit Currygewürzen an, bis es fast durch war, und fügte das Gemüse – Zwiebeln, Kartoffeln, Möhren – sowie Wasser und Salz hinzu. Er roch das intensive Aroma des brutzelnden Fleisches und der Kräuter, das sich im ganzen Gebäude entfaltete, und würzte sein kulina-

tion, curry in Korea was commercialised as an instant foodstuff, popular in homes, school canteens, hospitals and the armed forces. What you are going to eat tonight is this Korean variant of curry, which landed in South Korea after a journey from India via Great Britain and Japan, and which uses cornflour to dilute the authentic taste.

But because this answer is too complex and long-winded, I usually keep my response brief: "Korean-style curry."

Chapati

Starting from the shifting relevance and global compositions of meals, the following addresses the subject of food as a means of social interaction. The Greek philosopher Socrates (470–399 BCE) appreciated "good meals", and knew already their capacity of bringing "good people" together. But what is the significance of shared meals in asylum reception centres, facilities that are characterised by heteronomy and shared by people of different backgrounds involuntarily and in forced proximity to each other? This section examines how sharing meals can bring people together in reception centres.

It was the winter of 2018–2019 in Friedland Transit Camp. Arif Ibrahim made *dal*, a lentil stew, or curries of potatoes and vegetables. His friend Raafe prepared *chapati*, an unleavened flatbread, while Haani usually prepared the room for dinner and did the dishes afterwards. Arif seared the chicken in

Abb. 7: Zubereitung von Chapati in der Gemeinschaftsküche des GDL Friedland, 2019. ©Friedemann Yi-Neumann

Image 7: Preparing chapatis in a shared kitchen, in Friedland Transit Camp, 2019. ©Friedemann Yi-Neumann

risches Kunstwerk, bis er zufrieden war. »Ich könnte es besser machen, wenn ich die richtigen Küchenutensilien hätte«, sagte er, »aber es schmeckt trotzdem sehr gut.«

Obwohl die Zutaten an diesem Abend gut waren, bedurfte das Kochen in einer Gemeinschaftsküche eines Lagers der Improvisations- und Kompromissfähigkeit. Die Zubereitung der *Chapatis* zur Zufriedenheit aller drei stellte eine besondere Herausforderung dar, angesichts knapper finanzieller Ressourcen, um die notwendigen Zutaten zu kaufen, und mangelnder Küchenausstattung, etwa um *Chapati* zu backen. Raafe rollte den Teig mit einer Dose Rasierschaum aus (weil ein Nudelholz zu teuer war) – was nicht als Problem wahrgenommen wurde. Allerdings stellte sich heraus, dass es schwierig war, das Brot, das sonst über einem offenen Feuer gebacken

the curry spices until it was almost well-done, before adding the other vegetable ingredients – onions, potatoes and carrots – water and salt. He smelled the intense aroma of the meat and herbs flash-frying, which unfolded throughout the whole building, and seasoned his culinary artwork until he was satisfied. "I could do better if I had proper cooking utensils", he said, "but this is very yummy, nonetheless."

Even though the ingredients were good this evening, cooking in the shared kitchen of a camp requires improvisation and compromise. In particular, making *chapati* whose taste and appearance were acceptable to the three friends turned out to be a genuine challenge, because of the lack of money to buy the ingredients desired and inadequate cooking equipment.

wurde, auf einem Elektroherd zuzubereiten, da dieser nicht heiß genug wurde, um das Brot aufgehen zu lassen. Raafe legte es daher direkt auf die Herdplatte.

Trotz der Mängel, der Herausforderung und der sozialen und kulturellen Unterschiede bieten gemeinsame Mahlzeiten eine Gelegenheit, Nachbar:innen und Freund:innen einzuladen und zu unterhalten, selbst in den beengten Räumlichkeiten des Durchgangslagers. Neben dem Teetrinken ist dies ein zentrales Ereignis des Genusses und der vorübergehenden Gemeinsamkeit; oft vermag es, religiöse, ethnische und nationale Zugehörigkeiten zu überbrücken.

An diesem Abend kamen ein Afghane und ein Rumäne in Raafes und Haanis Raum – eigentlich für einen letzten kurzen

Raafe rolled out the dough with a can of shaving foam, since rolling pins were too expensive, and this was not a problem. However, browning the bread on an electronic stove – a step normally undertaken over an open fire – proved difficult, because the stove was not always hot enough to blister the crust. Raafe tackled this challenge by placing the flatbread directly on the stove.

Despite shortages and challenges, and despite the social and cultural differences between those dining, shared meals are an opportunity to invite and entertain neighbours and friends, even within the confined rooms of a transit camp. Alongside drinking tea, this is one of the central events of enjoyment and of tempo-

Abb. 8: Bei der *Chapati*-Zubereitung in der Gemeinschaftsküche des GDL Friedland, 2019. ©Friedemann Yi-Neumann

Image 8: Preparing *chapatis* in a shared kitchen, in Friedland Transit Camp, 2019. ©Friedemann Yi-Neumann

Foodways

Besuch, bevor Arif, Raafe und Haani in eine andere Folgeunterkunft verlegt wurden. Letztendlich blieben sie den ganzen Abend und genossen die Gastfreundlichkeit der drei. Die meisten der im Raum versammelten Menschen hatten lange Asylbiografien in verschiedenen Ländern; für zwei von ihnen war es sogar der zweite Aufenthalt im Durchgangslager Friedland. Sie sprachen miteinander auf Urdu, Deutsch, Englisch und Arabisch. Aufgrund der Sprachbarrieren war immer eine Person von den Gesprächen ausgeschlossen – bis zur nächsten Sprache gewechselt wurde und eine andere Person für eine Weile ›raus‹ war.

Ein Thema dieser Gespräche war der Vergleich von unterschiedlichen Speisen, Geschmäckern und Formen der Zubereitung, die sie von Zuhause kannten. »In Bagdad lassen sie den Fladenbrotteig für ein paar Tage gehen. Sie kneten ihn viele Male, bevor er gebacken wird, erzählte mir ein irakischer Bruder. Wir machen das etwas einfacher und schneller, und es schmeckt auch gut«, sagte Raafe, wobei er sich auf eine Unterhaltung bezog, die er in der Küche gehabt hatte, und jeder stimmte zu. Das Essen wurde immer wieder gelobt.

Diese Weggefährten und Nachbarn unterschiedlicher nationaler, ethnischer und religiöser Herkunft verhandelten das, was Tilmann Heil als »Minimalkonsens« bezeichnet und ein Zusammenleben trotz divergierender Ansichten, Identifikationen und Werte ermöglicht (Heil 2015: 317). »Es ist egal, ob jemand Muslim, Christ oder Jude ist oder sogar schwul. Hier vertragen wir uns alle und respektieren einander.« An jenem Abend wurden diese Sätze, die das Fundament der rary commensality, which are often able to bridge gaps between religious, ethnic and national affiliations.

On that evening, an Afghan and a Romanian came to Raafe and Haani's room, initially for a short last visit, as Arif, Raafe and Haani were to be transferred to another accommodation centre on the outskirts of Braunschweig, Germany. But the guests ended up staying all evening, enjoying the hospitality. Most of the people gathering in the room had long asylum-seeking biographies in different countries; in fact, for two of them, it was their second time in the Friedland transit camp. They talked with each other in Urdu, German, English and Arabic. Because none of those present spoke all four of these languages, there was always one person who could not participate in the conversation until it shifted to the next language, when it was then someone else's turn to be 'out' for a while.

One form their conversation took was comparing different dishes, tastes and forms of food preparation from home. "In Baghdad, they let the flatbread dough rise for several days. They knead it many times before baking – or at least that's what an Iraqi brother told me. Our method is simpler and faster, and it's good too," Raafe said, referring to an exchange he had had in the kitchen – and everyone agreed. The meal was praised repeatedly.

These companions and neighbours of different national, ethnic and religious backgrounds negotiated what Tilmann Heil calls "minimal consensus", which allows people to live together despite diverging views, identifications and values (Heil

Koexistenz ausdrücken, wie ein Mantra in unterschiedlichen Formen und Sprachen wiederholt.

Trotz solcher Zusammenkünfte und Gemeinsamkeiten, wie sie im vorangegangenen Bericht dargestellt wurden, sind Flüchtlingslager und Aufnahmezentren als Orte konzipiert worden, an denen Rechte und soziale Bindungen aufgelöst werden. Wie der italienische Philosoph Giorgio Agamben argumentiert, seien Lager eine Materialisierung des Ausnahmezustands und reduzierten die Existenz der Menschen auf das »nackte Leben« (Agamben 2002). Diese Aussage ist richtig, da der Ausnahmezustand von Lagern häufig soziale Bindungen und Rechte beschneidet. Die Aussage ist aber auch falsch, denn viele Ethnolog:innen haben gezeigt, dass solche Einrichtungen zugleich ambivalente Räume sozialer Interaktion sind (Scott-Smith und Breeze 2020).

In der folgenden Geschichte geht es um die Schwierigkeiten einer iranischen Familie, die versuchte, ihre Unterkunft für einige Zeit in ein Zuhause zu verwandeln, indem sie Gäste für gemeinsame Mahlzeiten einlud. Die Erzählung verdeutlicht die unterschiedlichen Auffassungen davon, was ein ›minimales Zuhause‹ ausmacht. Dieses Thema tauchte während der Forschung auf, als Ärgernisse über einige ganz alltägliche Hilfsmittel entstanden, was im folgenden Abschnitt veranschaulicht wird.

2015: 317). "It doesn't matter if someone is Muslim, Christian, Jew, or even gay. Here we get along, and respect each other." Like a mantra, these sentences were repeated in different versions and languages that evening, expressing the bedrock of coexisting in camps and of commensalities established – for the time being at least.

Despite such gatherings and commensalities presented in the previous account, refugee camps and reception centres have been conceptualised as places where rights and social bonds are dissolved. As the Italian philosopher Giorgio Agamben has argued, camps are a materialisation of the state of exception and reduce peoples' existences to 'bare lives' (Agamben 1998). This statement is correct since the state of exception prevailing in camps often curtails social ties and rights. However, the statement is also wrong, since many anthropologists have shown that such facilities are ambivalent spaces of social interaction (Scott-Smith and Breeze 2020). The following story addresses the difficulties of an Iranian family in turning their shelter into a home for a period of time by sharing food in a variety of hospitable acts. The piece highlights diverging understandings of what a 'minimal home' comprises. This issue emerged after irritations concerning some mundane and quotidian tools during my research, as the following passage illustrates.

Abb. 9 Geschirr, 2019. ©Friedemann Yi-Neumann

Image 9: Tableware, 2019. ©Friedemann Yi-Neumann

Geschirr und Fremdbestimmung

In Durchgangslagern prallen sehr unterschiedliche Konzepte, Vorstellungen, Praktiken und Materialitäten aufeinander, wenn es darum geht, was unter einer würdevollen und angemessenen Unterbringung verstanden wird. Um das durchaus spannungsvolle Verhältnis von Institutionen und Kulturen des Zuhauses zu verstehen, lohnt sich ein Blick auf Geschirr.

Im Frühjar 2019 hat mich meine Kollegin Hatice Pınar Şenoğuz mit einem iranischen Ehepaar und ihrer Tochter bekannt gemacht, die schon in mehreren Durchgangslagern untergebracht waren und nun gemeinsam in

Tableware and Heteronomy

In transit camps, very different concepts, ideas, practices, and materialities collide regarding what different individuals and groups understand to be dignified and appropriate accommodation. Why is it worth looking at tableware in order to understand the tense relationship between institutions and cultures of the home?

In spring 2019, my colleague H. Pınar Şenoğuz introduced me to an Iranian couple and their daughter, who were living together in a follow-up accommodation in Hannover, after having passed through

einer Folgeunterkunft in Hannover leben. Als Familie aus der gebildeten Mittelschicht, die seit mehr als einem Jahr unterwegs ist, leiden sie unter dem Mangel an Privatsphäre und den fehlenden Möglichkeiten, richtige häusliche und familiäre Praktiken zu etablieren, wie das angemessene Empfangen von Gästen.

Für sie gipfelte ihr Unverständnis der unflexiblen administrativen deutschen Vorstellung von einem Zuhause in dem Moment, in dem das Personal Geschirr in der Folgeunterkunft verteilte: drei Teller, drei Tassen, drei Bestecksätze – eines für jede Person. Empört nahmen sie ihr Geschirr entgegen; es schien für sie eine grundlegend falsche Vorstellung von einem Zuhause vorzuliegen. Ein elementarer Aspekt für sie, sich zu Hause zu fühlen, ist die Möglichkeit, Freunde einzuladen und von Freunden eingeladen zu werden. Aber wie können Leute eingeladen werden, wenn sie nicht bewirtet werden können? Geschirr für drei Personen ist dafür ganz offensichtlich nicht ausreichend.

Auch wenn sie es sehr schätzten, in Deutschland zu sein, waren die Auseinandersetzungen um das Geschirr und die Unmöglichkeit, auch nur um einen einzigen Gegenstand mehr zu feilschen als das Personal offiziell vorgesehen hatte, für sie ein Beweis dafür, dass die deutsche Bürokratie nicht willens oder nicht in der Lage sei, zu erkennen, wie wichtig solche Aspekte für andere Kulturen seien. Die »Wiedererlangung von Würde«, wie sie es nannten, und ein gewisses Maß an selbstbestimmtem, sozialem und familiärem Leben ist ohne entsprechendes Geschirr nicht möglich.

An solchen Kleinigkeiten lassen sich ›homing‹-Praktiken aufzeigen, die oft in einem spannungsreichen Verhältnis zu Lagerregeln

several transit camps. As a family from an educated middle-class background on the move for more than a year, they had been suffering from the lack of privacy and from the impossibility of establishing proper domestic and family practices, including the appropriate reception of guests.

For this Iranian family, their unawareness and incomprehension concerning the rigidly administrative German concept of home cultures culminated in the moment in which staff handed out tableware in their follow-up accommodation: three dishes, three mugs and three cutlery sets – one for each person. They were indignant at this allotment of dishes, which seemed to them like a crude misconception of what home actually is. For the Iranians, being able to invite friends and to be invited oneself is a fundamental part of feeling at home. But how can people be invited if there is no spare crockery to serve them on? Dishes for only three people are obviously not enough.

Even though they very much appreciated being in Germany, the quarrels over tableware and the impossibility of negotiating to get even a single item more than the staff had officially designated was proof for the family that German bureaucracy was unwilling or unable to recognize how significant such issues are for other (home) cultures. "Regaining dignity", as they called it, and a certain degree of self-determined social and family life is not possible, in this view, without appropriate tableware.

Such minor details reveal a lot about domestic practices, which are often enacted in a relationship of friction with camp

Foodways

Abb. 10: Foto des Geschirrs, das bei einem Folgebesuch in Hannover verwendet wurde, 2019. ©Friedemann Yi-Neumann

Image 10: Photo of the tableware used during the follow-up visit in Hannover, 2019. ©Friedemann Yi-Neumann

stehen. Dieses durchaus ungleiche Verhältnis möchte ich mit dem in Verbindung bringen, was Erving Goffman die »Diskulturation« (2014 [1961]: 24) totaler Institutionen nennt. Damit meint er den Verlust und einen Prozess des Verlernens üblicher Gewohnheiten, die normalerweise für das alltägliche Leben notwendig sind, aber in diesem Zusammenhang die Organisationsprozesse von Institutionen stören. Zu viele soziale Beziehungen und Zusammenkünfte in einem Aufnahmezentrum können für die Organisation der Lager eine Herausforderung darstellen. Deshalb registrieren sie häufig Besucher:innen und bitten diese, nach Ablauf der Besuchszeit zu gehen. Diese Form der Kontrolle und des Zwangs findet ihren Ausdruck in den begrenzten Zeitrahmen und auch der knappen Geschirrausstattung, die den Bewohner:innen gewährt werden.

rules. I propose a link between this quite unequal relationship to what Erving Goffman called the "disculturation" (1961: 73) of total institutions. By this he means the loss and "untraining" (ibid., p. 13) of habits that are usually necessary for everyday life but that in the new context, disturb the organisational processes of institutions. Having too many social engagements and events in a reception centre can challenge, and even apparently overchallenge, centre management organisations. This is evidenced by organisations requesting that visitors register and leave when the visiting hours are over, while in other migrant housing locations visits are not permitted at all. This form of control and constraint is also expressed in the limited timeslots allocated to visitors and the paucity of the dining equipment granted to the accommodation of residents.

Even though the cultural and organisational rules and norms imposed by reception authorities are dominant, they are in no way complete. People develop counterstrategies against institutionalised "disculturation", despite everything. The Iranian family, for instance, now borrows

Auch wenn die von den Aufnahmebehörden auferlegten kulturellen und organisatorischen Regeln und Normen vorherrschend sind, decken sie nicht alles ab. Menschen entwickeln Strategien gegen die institutionalisierte »Diskulturation«. Die iranische Familie beispielsweise leiht sich nun Geschirr von Nachbar:innen, um Gäste zu empfangen, was für sie, wie sie schildern, lästig und peinlich ist.

Solche kleinen Teilstücke materieller Wohnkulturen wie Geschirr können die Funktionsweisen von Lagern ans Licht bringen sowie auch die unterschiedlichen Perspektiven darauf und Konzepte davon, was notwendig ist und was es braucht, um »in Würde zu leben«, wie es die Mutter der iranischen Familie ausdrückte.

Wie die hier versammelten Beispiele zeigen, sind Essen und Migration nicht voneinander zu trennen, weder auf der historischen noch der aktuellen global-ökonomischen Ebene, noch was die Produktion von Nahrung und deren gesellschaftliche Folgen angeht. Essen kann wiederum ein wichtiger Bestandteil der Neuaushandlung und Neuverortung von Lebenswelten von Geflüchteten und Migrant:innen sein. Wo es die finanzielle und räumliche Situation erlaubt, ermöglicht es geschmackliche Bezüge zu Vergangenem und Vertrautem. Gleichzeitig verändert sich Essen in der Migration oder im Exil und kann die materielle Basis eines neuen Miteinanders und neuer Beziehungen werden. Umgekehrt kann Essen Fremdbestimmung und soziale Isolation materialisieren, wird aber auch zur Projektionsfläche von Ablehnung, Rassismen und Stereotypisierungen.

Der Anthropologe Michael Herzfeld hat diese soziale Komplexität des Essens unter dem Begriff der »food stereotypes« folgendermaßen zusammengefasst:

tableware from neighbours so that they can receive guests – "annoying and embarrassing" is how they describe this coerced dependency.

Tableware and other small puzzle-pieces of material home cultures can bring camp-operational modes to light and also provide insights into different perspectives and understandings of what is 'necessary' and what is 'needed' to live in dignity, to paraphrase the mother in the Iranian family.

The examples gathered here underline that food and migration cannot be artificially separated from one another, either at a historical or at a contemporary, global economic level, or with regard to food production and its societal consequences. In turn, food can form an important component in the endeavours of refugees and migrants to renegotiate and relocate their lifeworlds. Where the financial and spatial situation of actors permits, food facilitates the connections of taste to the past and to what is familiar. Concurrently, food changes during migration and exile and can form the material basis of a new coexistence and of new relationships. Conversely, in other permutations, food can function as a materialisation of heteronomy and social isolation, and it can also be hijacked as a surface on which to project rejection, racism and the stereotyping of groups and individuals.

The anthropologist Michael Herzfeld summarises this social complexity that attaches itself to food by using the concept of "food stereotypes":

Foodways

> [...] [Essens-]Stereotypen folgen dem Muster aller anderen Arten der Stereotypisierung und dienen gleichzeitig politischen und interaktiven Zwecken. Die Beschäftigung mit Lebensmitteln außerhalb der Kontexte, in denen sie konsumiert und diskutiert werden, wie Restaurants, der häuslichen Umgebung und Akten der Gastfreundschaft, ignoriert ihre zweischneidige Rolle. Essen kann sozialen und kulturellen Wandel vermitteln, auch wenn und gerade weil es den Anspruch erhebt, eine uralte Tradition zu repräsentieren. (Herzfeld 2016: 43–44)

> [...] [F]ood stereotypes follow the pattern of all other idioms of stereotyping, and similarly serve political and interactional ends. To treat food outside its contexts of consumption and discussion – restaurants, domestic settings and acts of hospitality – is to ignore its double-edged role. Food can mediate social and cultural change, even as, and precisely because, it claims to represent age-old tradition. (Herzfeld 2016: 43–44)

Essen kann also gleichzeitig Tradition repräsentieren und im Wandel sein, es kann Einschluss oder Ausschluss materialisieren und es nimmt im Exil häufig einen veränderten Stellenwert ein, wie es etwa am Beispiel *Injera* gezeigt wurde. Doch nicht nur Menschen migrieren, sondern auch das Essen bewegt sich auf historischen Wegen, die oftmals in Vergessenheit geraten. Dabei kann eine Beschäftigung mit den (Handels-)Wegen des Essens auch Fragen und Probleme seiner globalen politischen Ökonomie zum Vorschein bringen. Der Satz, dass gutes Essen gute Menschen zusammenbringt, wie Sokrates es einst formulierte, mag angesichts dieser Verwerfungen fragwürdig erscheinen. Dennoch bleibt die Frage, was Menschen unter ›gutem Essen‹ verstehen und deren Zugang dazu eine wichtige Überlegung. Die vertraute Küche mag für manche Migrierten mit Heimweh und Wehmut verbunden sein, die Betrachtung der eigenen Migrationsgeschichte kann aber durchaus auch zu einem selbstironischen Verhältnis führen und muss nicht immer zum Schluss kommen, dass das Essen von ›Zuhause‹ das Beste sei.

Thus, food can represent tradition and transition simultaneously; it can materialise both inclusion and exclusion; and foods often acquire a different value through exile, as has been shown through the example of *injera*. But not only people but also food migrate along historical routes, along land, sea and flight paths that often get forgotten about. Engaging with the (trade) routes along which food is transported can also bring to the fore questions and problems pertaining to the global political economy of food. Placed alongside such questions, Socrates' axiom that good food brings good people together might seem objectionable. Yet the question of what "good food" is remains open, and access to this quantity is a vital consideration. While familiar cooking is connected to feelings of homesickness and longing for many migrants, reflections on one's own migration history can also lead to a form of intrapersonal relations characterised by an ability to laugh at oneself. They do not have to culminate in the conclusion that food from 'home' is necessarily best.

Literatur | References

Agamben, Giorgio. 1998. *Homo sacer: Sovereign Power and Bare Life*. Stanford.

Agamben, Giorgio. 2002. *Homo Sacer. Die Souveränität der Macht und das nackte Leben*. Frankfurt a.M..

Cheng, Acga; Sean Mayes, Gemedo Dalle, Sebsebe Demissew, Festo Massawe. 2017. Diversifying Crops for Food and Nutrition Security – A Case of *teff*. *Biological Review* 92 (1): 188–198. https://doi.org/10.1111/brv.12225

Heil, Tillmann. 2015. Conviviality. (Re)negotiating Minimal Consensus. In: Steven Vertovec: *Routledge International Handbook of Diversity Studies*, 317–324. London.

Herzfeld, Michael. 2016. Culinary Stereotypes: the Gustatory Politics of Gastro-Essentialism. In: J.A. Klein; J.L. Watson: *The Handbook of Food and Anthropology*, 43–44. London.

Povrzanović Frykman, Maja. 2018. Food as a Matter of Being: Experiential Continuity in Transnational Lives. In: D. Mata-Codesal, M. Abranches: *Food Parcels in International Migration: Intimate Connections*, 25–46. Cham.

Scott-Smith, Tom; Mark E. Breeze. 2020. *Structures of Protection? Rethinking Refugee Shelter*. Oxford & New York.

Verdasco, Andrea. 2022. Cooking ›Pocket Money‹: How Young Unaccompanied Refugees Create a Sense of Community and Familiarity at a Danish Asylum Center. In: Friedemann Yi-Neumann, Andrea Lauser, Antonie Fuhse, Peter Bräunlein: *Material Culture and (Forced) Migration: Materializing the Transient*. London.

Zelalem, Zecharias. "Special Edition: Teffgate – A Dutchman's Conning of the Ethiopian State." *Addis Standard*, 27 November 2019, https://addisstandard.com/special-edition-teffgate-a-dutchmans-conning-of-the-ethiopian-state/

Anmerkungen

1. Weltweit arbeiten Migrant*innen häufig in der Landwirtschaft, also der Nahrungsmittelproduktion. Gerade im Kontext der COVID-19-Pandemie sind diese besonders großen gesundheitlichen und ökonomischen Risiken ausgesetzt. Siehe dazu: Food and Agriculture Organization of the United Nations. 2020. *Migrant Workers and the COVID-19 Pandemic*, Rome 2020. https://doi.org/10.4060/ca8559en
2. Eine vergleichende Studie zu diesem Thema: Bonfanti, Sara; Aurora Massa; Alejandro Miranda Nieto: »Whiffs of Home. Ethnographic Comparison in a Collaborative Research Study Across European Cities«. *Etnografia e ricerca qualitative*, 2, 2019, S. 153–174. Zu Rassismus und *Berbere* siehe S. 164.
3. 2012 wurde der Deal kritisiert sowie dessen Folgen für die Ernährung, Ökonomie und Biodiversität herausgearbeitet. Siehe dazu Regine Andersen; Tone Winge: »The Access and Benefit-Sharing Agreement on Teff Genetic Resources – Facts and Lessons«, *FNI Report* 6/2012 (http://www.abs-initiative.info/fileadmin/media/Knowledge_Center/Pulications/FNI/FNI-R0612.pdf).
4. Siehe Patenteintrag: https://patents.justia.com/inventor/jans-roosjen

Notes

1. Across the globe, migrants often work in agriculture – i.e., in food production. Particularly in the context of the Covid-19 pandemic, these workers face especially grave risks, in both health terms and economically. See Food and Agriculture Organization of the United Nations: *Migrant Workers and the COVID-19 Pandemic*, Rome 2020. https://doi.org/10.4060/ca8559en
2. For a comparative study on this subject, see: Bonfanti, Sara; Massa, Aurora & Alejandro Miranda Nieto. (2019:) "Whiffs of Home. Ethnographic Comparison in a Collaborative Research Study Across European Cities." *Etnografia e ricerca qualitative*, 2, p. 153–174. On racism and berbere see, pp. 164.
3. The deal itself, and the impact it has had and will have on nutrition, the economy and biodiversity, was criticised in 2012. See: Regine Andersen; Tone Winge: "The Access and Benefit-Sharing Agreement on Teff Genetic Resources – Facts and Lessons", *FNI Report* 6/2012, http://www.abs-initiative.info/fileadmin/media/Knowledge_Center/Pulications/FNI/FNI-R0612.pdf
4. See the list of patents filed with, and patents already granted by, the United States Patent and Trademark Office (USPTO): "Patents by Inventor Jans Roosjen", *Justia Patents*, accessed 4 November 2021, https://patents.justia.com/inventor/jans-roosjen
5. *Geduldete,* which translates literally as "tolerated persons", is a formal category in German Residency Law. It is used to describe those non-German residents with a legal obligation to leave the country whose deportation has been "temporarily suspended", often because the country/countries these individuals could be deported to have been officially categorised as unsafe. Such "temporary suspensions" of deportations can last 20 years or more, and the category is usually imposed on those asylum-seekers and other migrants whose applications for a better residency status under German Asylum Law have been rejected. Freedom of movement is often restricted in Germany for *Geduldete*, as is access to work permits.

Schlüssel

Keys

Elza Czarnowski

Menschen auf der Flucht tragen persönliche Besitztümer mit sich, teilweise unter großen Schwierigkeiten. Diese Dinge dokumentieren die Widerstandsfähigkeit und den Einfallsreichtum von Menschen, die vieles verloren haben (Auslander und Zahra 2018: 1f.). Sie bieten, außerhalb ihres eigentlichen funktionalen Kontextes, eine Möglichkeit, sich zu erinnern, sich verbunden zu fühlen und sich seiner selbst zu versichern. Außerdem können sie helfen, die Hoffnung (auf ein normales Leben, auf Rückkehr, auf ein Wiedersehen) zu bewahren. Die Dinge verändern sich und ihre Wirkung auf ihre Träger:innen, sie besitzen also ein Eigenleben. Bei manchen Gegenständen ist die Transformation ihrer Bedeutung und die Wirkung auf die Menschen, denen sie gehören, besonders offensichtlich – so etwa bei dem alltäglichsten und unscheinbarsten darunter: dem Haustürschlüssel.

Ein Schlüssel. Ein meist kleiner, funktionaler Gegenstand, allgegenwärtig. Ein Stück Metall, das überall auf der Welt ähnlich aussieht. Viele Menschen besitzen einen Schlüssel – oftmals auch mehrere – und nutzen diese(n) tagtäglich, ohne dem Ganzen eine größere Bedeutung beizumessen. Erst wenn der Alltag zur Ausnahmesituation wird, wird das Augenmerk auf die kleinen, scheinbar

People fleeing threatening realities carry personal belongings with them, even when doing so causes them great difficulties. These things document the resilience and inventiveness of people who have lost a lot (Auslander and Zahra 2018: 1f.). Aside from their everyday function, they offer a way of remembering, of feeling connected and of reassuring their bearers. Moreover, they can help people to keep hoping – for a normal life again, for finding a way to return or for a reunification with loved ones. These carried things undergo change, both in themselves and in the effects they have on their carriers, seemingly demonstrating a life of their own.

The transformation in the significance of some objects is especially visible, as is their influence on the individuals to whom they belong. Like the house-key, for example, one of the most everyday and seemingly unremarkable possessions that migrate along with their owners.

A key – a usually small, ubiquitous, functional object. A piece of metal that looks similar everywhere on the planet. Many people possess a key – many also possess several – and use this or these key(s) day in day out, without attaching any

ausschließlich zweckgebundenen Dinge gelenkt, die das Gewöhnliche in Außergewöhnliches verwandeln (ebd.: 3).

Was also ist ein Schlüssel, welche Funktion schreiben wir ihm zu – und wie vermag sich diese zu verändern, wenn sich die Rahmenbedingungen des Lebens der Menschen ändern, die diesen Schlüssel besitzen?

Die Hauptfunktion eines Schlüssels ist klar: Er soll etwas öffnen – oder verschließen. Er bietet also Sicherheit und Schutz vor unbefugtem Betreten, Benutzen oder anderweitigem Zugriff und verleiht seinen Träger:innen so die Möglichkeit, für sie Wertvolles zu verwalten. Ganz eng daran knüpft sich die symbolische Hauptfunktion eines Schlüssels als legitimierendes Zeichen von Macht und Kontrolle über einen ab- bzw. eingeschlossenen Bereich: Nur wer einen passenden Schlüssel besitzt, hat die Macht und das Recht, eine bestimmte Tür zu öffnen.

In der christlichen Symbolik verdeutlicht dies zum Beispiel eine Bibelstelle im Neuen Testament, in der Jesus die Schlüssel des Himmelreiches dem Jünger Petrus übergibt und ihm so symbolisch die sakramentale Macht des ›Bindens‹ und ›Lösens‹ verleiht[1]: Petrus erhält mit diesem Schlüssel die Aufgabe, regierende Verantwortung auf der Erde zu übernehmen und damit einhergehend die Autorität, Sünder:innen zu benennen (›binden‹) und deren Sünden zu vergeben (›lösen‹).

Die symbolische Bedeutung von Schlüsseln findet sich auch bei kriegerischen Auseinandersetzungen: Seit dem Mittelalter war eine zeremonielle Überreichung der Schlüssel eroberter Städte und Festungen an die

larger significance to the larger procedure. Only when everyday life itself becomes an exceptional situation is our attention redirected to these small things, which seem exclusively tied to a single purpose: when the ordinary becomes unordinary (Ibid. 3).

What, then, is a key? Which functions do we attribute to it, and how do these change when the conditions alter that structure the lives of the keys' owners?

The main function of the key is evident. It is meant to open something – or to lock it. It provides security and protection against unauthorized entry, use or any form of access, and it grants to those who carry it the possibility of administering something valuable. Closely tied to this is the principal symbolic function of the key as a legitimizing sign of power and control over some specified locked or enclosed space: Only those possessing a key that fits have the power and legal right to open a particular door.

The Christian symbolism of keys is illustrated and exemplified by a Bible passage from the New Testament, in which Jesus gives his disciple Peter the keys to the kingdom of heaven, thereby symbolically granting him the sacramental power of 'binding' and 'loosing' or, in modern language, of releasing.[1] Along with these keys, Peter is given the responsibility for governing the earth, and with that the authority to name sinners – to 'bind' them – and to forgive their sins – by 'loosing' or releasing sinners from the same.

Keys as symbolically significant objects are also prominent in the history of military conflicts: In the Middle Ages, the

Sieger:innen als Ausdruck der Machtübergabe üblich.

Diese beiden Beispiele für die Symbolkraft von Schlüsseln wurden von europäischen Künstler:innen aufgegriffen und bildeten über Jahrhunderte und Epochen hinweg ein beliebtes Motiv, was die symbolische Funktion nicht nur reproduzierte, sondern durch Wiederholung sogar verstärkte.

Doch in einer Zeit, in der Schlüssel und Schließmechanismen zum einen leicht zu kopieren sind und zum anderen die Ausübung von Gewalt Menschen dazu zwingt, Türen hinter sich zu verschließen und Orte zu verlassen, an denen Türen und Häuser mit anderen Mitteln geöffnet oder gar zerstört werden, kommen weitere Bedeutungsebenen hinzu – und die ursprüngliche Funktion tritt mehr und mehr in den Hintergrund.

So kann der Haustürschlüssel in den Händen einer geflüchteten Person nicht nur Erinnerungen hervorrufen – an das eigene Haus, die Wohnung, die Menschen darin, an besondere Momente und alltägliche Belanglosigkeiten, an liebgewonnene Gewohnheiten, an Sorgen, Gefühle, Gedanken und Wünsche: also an das Leben vor der Flucht. Diesem Schlüssel mag, als biografisches Objekt, auch eine selbstvergewissernde und stärkende Wirkung innewohnen: Dieser eine Schlüssel zu diesem einen Schloss macht seine:n Träger:in zu einer Person, die eine Heimat hat, eine eigene Geschichte an einem ganz bestimmten Ort, ein selbstbestimmtes Leben vor der Gegenwart. Eine Aufgabe, einen Beruf, Pläne. Eine Identität, die sich nicht aus den bürokratischen Verfahren einer sogenannten Aufnahmegesellschaft und deren Rollenzuweisungen speist. So wird beispiels-

ceremonial handing over of the keys to conquered cities and fortresses to the victors became a traditional way to express the handover of power.

Both examples of the symbolic power of keys were taken up by European artists and functioned as a favourite motif down the centuries and changing epochs. This representation didn't merely reproduce the symbolic function of keys, it even strengthened it.

Yet, we live in an age in which keys and locks can be copied easily, and in which violence forces people to lock their doors behind them for good. This same violence pushes people to abandon localities in which doors and houses can be forcibly opened again by other means, or even destroyed completely. These contemporary realities lend keys and locks further layers of significance – moving their original function further and further into the background.

A frontdoor key, in the hands of a refugee, is more than simply a conjuror of memories – of one's own house or apartment, of the people inside that home, of special moments and inconsequential occurrences, of familiar habits, of worries, feelings, thoughts and wishes; in short, of life before having to flee. Beyond this, a self-reassuring and reinforcing effect can also be immanent in a key as a biographical object. A particular key fitting into a particular lock turns its bearer into a person who has a home, a personal history at an utterly concrete place, a self-determined life before the present day. A task, a vocation – plans. An identity that cannot be

weise unter geflüchteten Palästinenser:innen der Schlüssel als Symbol ihres Rechts auf Rückkehr in ihre Häuser, in die Heimat ihrer Vorfahr:innen, verstanden (Schiocchet 2014).

Natürlich ist die Gegenüberstellung von ›vor der Flucht – nach der Flucht‹ komplex und abhängig von jeder einzelnen Geschichte, jeder einzelnen Person und von gesellschaftlichen Hintergründen. Es ist zu vereinfachend, eine Bewertung anhand der linearen Zuschreibung ›vorher – nachher‹ vorzunehmen. Nicht alle Menschen, die geflüchtet sind, waren zuvor glücklich. Nicht jeder Schlüssel steht für ein selbstbestimmtes Leben in der Vergangenheit.

Neue Schlüssel zu einem neuen Heim ersetzen nicht immer die alten. Die Ethnologin Maja Povrzanović Frykman, die an zwei Orten ein Zuhause hat – in Schweden und Kroatien – beschreibt ihre Erfahrung mit ihren beiden Haustürschlüsseln: »Ich trage die Schlüssel zu meinen beiden *homes* auf meinen Reisen in beide Richtungen immer bei mir [...]. Mir ist bewusst, dass ich sie mitnehme, weil sie mir die praktische Möglichkeit geben, mein jeweiliges Zuhause auf die ›normale‹ Art und Weise zu betreten« (2016: 45, Übersetzung AF). In ihrem ersten Jahr in Schweden wurde sie allerdings vom vermeintlich selbstverständlichen Akt des Türöffnens ständig an ihre Fremdheit erinnert: »Jeden Tag, immer und immer wieder, drehte ich den Schlüssel in die Richtung, die für Schlüssellöcher in Kroatien Standard ist und wurde dann daran erinnert, dass es in Schweden die falsche ist« (ebd.).

Die praktische Nutzung von Alltagsgegenständen wie die eines Schlüssels in einem

reduced to mere bureaucratic procedures, as they manifest themselves in so-called receiving societies. Palestinians in the diaspora, for example, understand keys as a symbol for their right to return to their houses in the home country of their ancestors (Schiocchet 2014: 130–60).

The dichotomy of 'before flight – after flight' is evidently complex and is dependent on each individual story and person, and the societal backgrounds from which these individuals came. To evaluate this dichotomy based merely on the linear attribution of 'before and after' would be a simplification. Not all humans were happy before they fled. And not all keys represent an autonomous past life.

New keys to a new home don't necessarily replace the old ones. The ethnologist Maja Povrzanović Frykman, who has a home in two localities, in Sweden and in Croatia, describes how she has experienced having two front-door keys: "The keys to my two homes always travel with me in both directions … I am aware that I carry them as providers of the immediate and practical possibility of re-entering my respective homes in a 'normal' way" (2016: 45). However, Frykman was constantly reminded of her foreignness during the first year in Sweden through the supposedly automatic action of unlocking a door: "Every day, over again and over again, I turned the key in the direction normal for all keyholes in Croatia, only to be reminded that here, in Sweden, it is the wrong direction" (ibid.).

The practical use of everyday objects, like a key, in a new context may remind

neuen Kontext kann die eigene Fremdheit in Erinnerung rufen oder auch eine optimistische und sinnstiftende Wirkung entfalten. Das Betrachten dieses Objekts vermag einem Gefühl der Ohnmacht und Ausweglosigkeit etwas entgegenzusetzen: nicht nur die konkrete Erinnerung an die Vergangenheit, sondern die Vergewisserung, dass ein einzelner Mensch etwas bewirken kann – für sich und andere.

Die Journalisten Christoph Cadenbach, Lars Reichardt und Alexander Gehring (Fotos) haben 2018 für ihren Beitrag »Schlüssel aus der Heimat« im *SZ-Magazin* mit geflüchteten Menschen gesprochen, die ihre Schlüssel nach Deutschland mitgebracht haben (Cadenbach et al. 2018). Dabei entstanden die folgenden sechs Geschichten: Portraits über Menschen, ausgehend von einem kleinen Gegenstand. Sie erzählen einerseits von Erinnerungen und andererseits von Gefühlen wie der Trauer über den Verlust, der Wut über das Geschehene, der Hoffnung auf Rückkehr und der Erleichterung, am Leben zu sein.

someone of their own foreignness, but it may also unleash an optimistic or meaningful effect. Observing such an object could set up a barrier against feelings of powerlessness or of having no way out of a particular situation. This goes beyond concrete memories of the past to a feeling of being reassured that a single person can have an impact – on their own lives and on the lives of others.

For their 2018 article "Keys from Home" in the magazine of the *Süddeutsche Zeitung* [a southern German newspaper], accompanied by the photographer Alexander Gehring, the journalists Christoph Cadenbach and Lars Reichardt spoke with refugees who had brought their keys along with them to Germany (Cadenbach et al. 2018).

The six stories that follow emerged during the work on the article: portraits of people with a small object as their point of departure. They communicate memories but also feelings, including grief about what has been lost, fury about what has happened, hopes of return and a sense of relief to still be alive.

Abb. 1: Der Schlüssel von Ahmad M., 2018. ©Alexander Gehring

Image 1: The key of Ahmad M., 2018. ©Alexander Gehring

Schlüssel

Ahmad M., Aleppo, Syrien

Was wohl aus seinen Büchern geworden ist, die im Regal in seinem Zimmer standen und die er so geliebt hat? Ob der Nachbar, der nun ihre Wohnung besetzt, sie liest? Ob er sie weggeworfen hat?

Ahmad will sich das gar nicht vorstellen. In der Erinnerung, die er zulässt, sitzt er mit seinen fünf Geschwistern in der Wohnküche ihres Apartments in Aleppo, warmes Licht scheint durchs Fenster, es duftet nach Essen, das seine Mutter gekocht hat. Seine Eltern hatten die Wohnung 2005 gekauft: 150 Quadratmeter, dritter Stock, großer Balkon, gute Gegend von Aleppo. Seit 2014 lebt Ahmad mit seiner Familie in Berlin, im Wedding.

Ein Freund von ihm, der in Aleppo blieb, hat ihm erzählt, dass eine Bombe die Apartments in den oberen Stockwerken ihres Hauses zerstört hat. Und dass der Nachbar, den sie gebeten hatten, nach dem Rechten zu sehen, sich breitgemacht hat in ihrer Wohnung. Wegen der chaotischen Lage in Aleppo hindert ihn bisher niemand daran.

In Berlin macht Ahmad eine Ausbildung zum Großhandelskaufmann. In Syrien hatte er Tourismus studiert. Er ist 27. Den Schlüssel verwahrt er weiter, obwohl er vermutlich nutzlos ist: Der Nachbar soll das Schloss mittlerweile ausgetauscht haben.

Ahmad M., Aleppo, Syria

What has become of his beloved books that stood on the shelf in his room? Does the neighbour now occupying their flat read them – or has he thrown them away?

Ahmad doesn't want even to think about it. In the memories he does permit, he's sitting with his five siblings in the kitchen/living room in their Aleppo flat; warm light is streaming through the window, and the flavours of the food his mother has just cooked fill the air. His parents bought the flat in 2005: 150 square meters, third floor, large balcony, and in a good district of the city. Since 2014, Ahmad lives with his family in the Berlin district of Wedding.

A friend who remained in Aleppo told him that a bomb had destroyed the flats in the upper stories of their house. And that the neighbour, whom Ahmad's family had asked to keep an eye on the flat, had made himself at home there instead. The chaotic situation in the city means there is now no one there to block such unsanctioned steps.

In Berlin, Ahmad is training to work in wholesale; in Syria he had studied tourism. He still keeps his key safe, even though it feels useless now. Ahmad was told that the neighbour has since changed the lock.

Husain Alkhalil, Almahdum, Syrien

Sein Haus stand ganz in der Nähe der *muḥtasibūn*, der Personen, die die *hisba* durchsetzen, also die Einhaltung der Ordnung der Scharia. Sie hatten ihre örtliche Zentrale nur 200 Meter weiter eingerichtet, um von dort aus die Menschen im syrischen Dorf Almahdum, 70 Kilometer vor Aleppo, mit Stockschlägen gottesfürchtiger zu machen. Kein Mann sollte sich mehr den Bart rasieren und die Haare schneiden. Husain Alkhalil, 33, ging den *muḥtasibūn* aus dem Weg. Sein Haus geriet immer wieder unter den Beschuss von Regierungsflugzeugen. Husain Alkhalil meint, es müssten Flugzeuge unter Assads Kommando gewesen sein, die Amerikaner:innen hätten sicherlich besser gezielt und die Stellung des IS nicht verfehlt.

Alkhalil hatte als LKW-Fahrer in Saudi-Arabien gearbeitet. Der Job hatte ihm ermöglicht, Rücklagen zu bilden, sodass er alle 6 Monate für 2 Monate in die Heimat zurückkehrte. 2009, nach seiner Hochzeit, hatte er mit dem Hausbau begonnen, erst mal zwei Zimmer, ein Dach und ein Zaun drum herum. Später kamen zwei Schlafzimmer hinzu, der Hof wurde ummauert, 180 Quadratmeter alles in allem. Einen Schlüssel brauchte die Familie damals nicht. Erst als der IS kam, 2014, schlossen die 5.000 Bewohner:innen von Almahdum ihre Häuser ab. Eineinhalb Jahre stand das Dorf unter IS-Herrschaft, als das Haus von Alkhalils Bruder bombardiert wurde. Dabei starben Husain Alkhalils Schwägerin und sein Neffe, seine siebenjährige Nichte verlor ein Bein. 2015 war dann Husain Alkhalils Haus dran:

Husain Alkhalil, Almahdum, Syria

His house stood very near to the headquarters of the *muḥtasibūn*, a traditional term for Islamic officials who enforce *Hisbah* – a synonym for Sharia law. Only 200 metres away from Husain's house, these individuals had set up their local headquarters as a base from which to make the people of Almanhdum – a village 70 kilometres from Aleppo – fear God more by beating them with sticks. A ban was put on village men shaving or cutting their hair. Husain Alkhalil, 33 years old, did his best to avoid the *muḥtasibūn*. On top of that, his house was coming under regular fire from government planes. Husain Alkhalil argues that these must have been planes under Assad's command, as the Americans with their better aim would have managed to hit the Islamic State's position.

Alkhalil used to work as a lorry driver in Saudi Arabia. The job enabled him to save money, so that every 6 months he returned home for a 2-month stay. After his wedding in 2009, he began to build a house in Syria: first simply two rooms with a roof and with a fence around it. Two further bedrooms were added later as well as a wall built around the compound – a total of 180 square meters of living space. At that time, the family didn't even need a key. It was only when the IS arrived in the village in 2014 that the 5,000 inhabitants of Almahdum started to lock their houses. The village had been under IS rule for around 18 months when Alkhalil's brother's house was bombed. Husain's sister-in-law and his

Abb. 2: Der Schlüssel von Husain Alkhalil, 2018.
©Alexander Gehring

Image 2: The key of Husain Alkhalil, 2018.
©Alexander Gehring

Abb. 3: Die Schlüssel von Janet Sadeq, 2018.
©Alexander Gehring

Image 3: The keys of Janet Sadeq, 2018.
©Alexander Gehring

Schlüssel

Alle Fenster gingen zu Bruch, Munitionssplitter blieben in der Wand stecken. Alkhalil floh mit Frau und drei Kindern erst in ein kurdisch besetztes Dorf, dann in die Türkei. Eine fremde Familie ließ über einen im Dorf zurückgebliebenen Onkel anfragen, ob sie in Alkhalils Haus wohnen dürften. Alkhalil riet ab, die Flugzeuge würden sicherlich wiederkommen. Zehn Tage später kamen sie – und zerstörten das Haus restlos.

Alkhalils Familie floh über die Balkanroute von der Türkei nach Berlin. Den Schlüssel möchte er nicht wegwerfen.

Janet Sadeq, Mossul, Irak

Der Schlüssel ist das Einzige, was Janet Sadeq von der 300 Quadratmeter großen Villa in Mossul und dem 500 Quadratmeter großen Garten mit vielen Blumen und etlichen Orangen- und Olivenbäumen geblieben ist. Janet Sadeqs Mann gehörte zur christlichen Minderheit, verkaufte Alkohol, seine Geschäfte liefen lange Zeit gut. Und Janet Sadeq, heute 54, arbeitete schon unter Saddam Hussein als gut bezahlte Krankenschwester. 1998 wurde ihrem Mann die Lizenz entzogen. 2005 kamen die Amerikaner:innen, landeten mit einem Hubschrauber im Garten und schossen um sich. Niemand kam zu Schaden, aber alle Fenster waren zerstört. Das Armeekommando suchte Terrorist:innen. Wahrscheinlich hatte ein Nachbar behauptet, die Familie Sadeq verstecke welche. Janets Mann starb 2013 an den Folgen eines Unfalls, 6 Jahre lang hatte sie ihn versorgt. Als der IS dann im Juni 2014 auf Mossul marschierte, waren die Sadeqs schon weg. Aus Furcht vor Ent-

nephew were killed outright, and his seven-year-old niece lost a leg. In 2015, it seemed that it was the turn for Husain Alkhalil's own house: The bombing broke all the windows, and ammunition splinters were left stuck in the walls. Alkhalil fled with his wife and three children first into a village occupied by Kurds and then on to Turkey. A family unknown to Alkhalil enquired via an uncle who had remained in the village whether they would be allowed to live in Alkhalil's house. Alkhalil advised against the move, maintaining that the planes were bound to return. And 10 days later they did – bombing the house flat.

Alkhalil's family fled via the established route through the Balkans – known in German simply as the 'Balkanroute' – from Turkey to Berlin. But Husain does not want to throw away the key to his Syrian house.

Janet Sadeq, Mosul, Iraq

The key is the only thing Janet Sadeq still has left from her large villa in Mosul, 300 m^2 in size, and her sizeable garden – 500 m^2 – with all its flowers, orange trees and olive trees. Janet Sadeq's husband belonged to the Christian minority and sold alcohol – and his business did well over a long period. For her part, Janet, who is 54 now, worked outside the home, during Saddam Hussein's presidency already as a well-paid nurse. But Janet's husband's license to sell alcohol was revoked in 1998. In 2005, the Americans arrived, landed in the garden with a helicopter and fired shots in all directions. No one was injured, but all the

Abb. 4: Der Schlüssel von Mohammad al Massad, 2018.
©Alexander Gehring

Image 4: The key of Mohammad al Massad, 2018.
©Alexander Gehring

Schlüssel

führungen waren eine Tochter und ein Sohn zuerst gegangen. Die Tochter fand Arbeit als Übersetzerin in Berlin und holte ihre Mutter Janet nach. Die Villa in Mossuls Norden, ganz in der Nähe der Universität, stand nicht lange leer. Die Soldaten des IS zogen ein – bis sie ausgebombt wurden.

Mohammad al Massad, Baiyt Irah, Syrien

Aus ihrem Haus ist ein Gefängnis geworden, eine Art Polizeistation der Terrororganisation Islamischer Staat: Das ist die letzte Information aus seinem Heimatdorf Baiyt Irah, die zu Mohammad al Massad nach Berlin durchgedrungen ist. Mohammad al Massad ist das älteste von vier Geschwistern, 22 Jahre alt. Als der IS in sein Dorf kam, waren er und seine Familie bereits in die nächste Stadt geflohen. Das war 2015. Er zog dann allein weiter in Richtung Europa.

Ihr Haus war umgeben von einem großzügigen Garten mit Dutzenden von Oliven- und Zitronenbäumen. Sie pressten ihr eigenes Öl, bauten Kartoffeln, Tomaten und Paprika an. Al Massads Vater arbeitete als Brunnenbauer. Al Massad will in Deutschland bald eine Ausbildung zum Brunnenbauer beginnen.

Der Schlüssel und ein Halstuch seiner Mutter sind die wichtigsten Dinge, die ihn in Berlin an seine Vergangenheit erinnern. Er vermisst den Blick aus dem Fenster seines früheren Zimmers, an dem er manchmal heimlich geraucht hat. Ein Blick auf fruchtbares, bergiges Land. Sein Heimatdorf liegt im äußersten Südwesten von Syrien, aus seinem

windows were broken. The army squadron claimed to be looking for terrorists, following a tip-off that appears to have come from one of the Sadeqs' neighbours. In 2013, Janet's husband died from the long-term effects of an accident, after she had cared for him for 6 years. By the time the Islamic State marched into Mosul in June 2014, the Sadeqs had already departed. Anxious about being kidnapped, a daughter and a son left already before the parents. The daughter found work as a translator in Berlin and arranged for her mother to follow after her. The villa in the north of Mosul, very close to the university, was not empty for long. IS soldiers moved in – until they, in turn, were bombed out of the place.

Mohammad al Massad, Baiyt Irah, Syria

The family house had been turned into a prison – or, more precisely, a kind of police station for the terror organisation known as the "Islamic State": That was the last news about his home village of Baiyt Irah that made its way through to Mohamad al Massad in Berlin. Mohamad, aged 22, is the eldest of four siblings. When the IS entered his village, he and his family had already fled to the nearest larger town. This first flight, in 2015, was followed by a further migration towards Europe.

The family's house was surrounded by a bountiful garden containing dozens of olive and lemon trees. They pressed their own oil and grew potatoes, tomatoes and peppers. Al Massad's father worked as a well-engi-

Fenster konnte er bis Jordanien und Israel sehen. Mohammad al Massad glaubt nicht, dass er noch einmal aus diesem Fenster blicken werde.

Hiba A., Latakia, Syrien

Der Schlüssel hängt an ihrem Schlüsselbund, den sie ständig bei sich trägt, obwohl sie ihn seit vier Jahren nicht mehr benutzt hat. Das Apartement, das er aufschließt, liegt im Zentrum von Latakia, einer syrischen Küstenstadt, auf die noch keine Bombe gefallen ist. Dort leben vor allem Menschen, die das Assad-Regime stützen. Hiba floh, weil sie Probleme mit diesen Menschen bekommen hatte. Mehr will sie dazu nicht sagen. Ihre Eltern leben immer noch dort, in diesem Apartement. Hiba, die heute 32 Jahre alt ist, hat in Latakia englische Literatur studiert und anschließend Arbeit gefunden, einen Verwaltungsjob im Gesundheitswesen. An den Wochenenden ging sie mit Freund:innen wandern oder an den Strand. Zu Hause saß sie gern auf dem sonnigen Balkon. In ihrer Erinnerung duftet das Apartment nach dem Jasmin-Parfüm ihrer Mutter. Allein das Wohnzimmer ist so groß wie ihre jetzige Wohnung in Berlin, in der sie mit ihrem Freund und ihrem 5 Monate alten Sohn lebt. Das Einhorn am Schlüsselbund gehört ihm.

Wenn Hiba den alten Schlüssel betrachtet, denkt sie oft an die Eltern einer palästinensischen Freundin, die 1948 nach Syrien flohen. Auch sie nahmen damals ihren Hausschlüssel mit und bewahrten ihn ihr Leben lang auf – ohne ihn je wieder benutzen zu können.

neer. Soon, Al Massad wants to start training as a well-engineer in Germany.

The key and his mother's shawl are the most important to objects remind him, in Berlin, of his past. He misses the view out of his old window, where he used to sometimes smoke secretly. It had a view over fertile, hilly country. His home village lies in southwest Syria, and out of his window he was able to see Jordan and Israel. Mohamad al Massad does not believe that he will ever look out of that window again.

Hiba A., Latakia, Syria

The key hangs from the key ring that she always carries with her, even though she hasn't used it for the last four years. The flat it unlocks is situated in the centre of Latakia, a Syrian coastal town on which to date no bombs have fallen. Most of its inhabitants support Assad's regime. Hiba fled because she had gotten into conflicts with these inhabitants – she doesn't want to talk about this in any depth. Her parents still live in the town – indeed in the same flat. Hila, who is 32 years old today, studied English literature in Latakia and found an administrative job in the healthcare system after her studies. On the weekends she liked to go walking with her friends or go to the beach. At home, she enjoyed sitting on her sunny balcony. In her memories, the flat smells of her mother's jasmine-scented perfume. Just the Syrian sitting room by itself is as big as her whole flat in Berlin now, in which she lives with her boyfriend and their 5-month-old son: It is his unicorn on the key ring.

Abb. 5 Der Schlüssel von Hiba A., 2018. ©Alexander Gehring   Image 5: The key of Hiba A., 2018. ©Alexander Gehring

Abb. 6: Die Schlüssel von Mohamad Riad al Aga, 2018.
©Alexander Gehring

Image 6: The keys of Mohamad Riad al Aga, 2018.
©Alexander Gehring

Schlüssel

Mohamad Riad al Aga, Aleppo, Syrien

Es war eine teure Gegend. Auf 400 Quadratmetern lebten Mohamad Riad Al Aga und seine Frau im fünften Stock eines Wohnhauses in Aleppo: Parkett, Kronleuchter, vier Bäder und vier Kinderzimmer, die Kinder waren schon erwachsen und ausgezogen, als der Bürgerkrieg nach Aleppo kam. Zwei Haushaltshilfen, eine aus Äthiopien, die andere von den Philippinen, gingen dem Ehepaar zur Hand. Al Aga, 64, hatte eine Fabrik für Rolltreppen, sieben Kilometer von der Wohnung entfernt. Er floh, als das Nachbarhaus in die Luft flog und eine Rebell:innengruppe begann, in seiner Fabrik Waffen zu produzieren. Er schloss seine Wohnung ab und vermauerte die Tür. Er hofft, dass das Haus noch steht, wenn Syrien irgendwann sicher genug ist, um zurückzukehren.

When Hiba reflects upon the old key, she often thinks of the parents of a female Palestinian friend, who fled to Syria in 1948. Back then, her friend's parents took their housekeys with them and kept them safe their whole life long – without ever being able to use them again.

Mohamad Riad al Aga, Aleppo, Syria

It was an expensive area. Mohamad Riad Al Aga and his wife had a full 400 m² at their disposal on the fifth floor of an apartment block in Aleppo: wooden-panelled floors, chandeliers, four bathrooms and four children's bedrooms – although the children were already grown up, and had already moved out when the civil war arrived in Aleppo. Two maids, one from Ethiopia, the other from the Philippines, lent the married couple a helping hand. Al Aga, now 64, had run an escalator factory 7 kilometres away from the flat. He fled when the neighbour's house was bombed out, and a group of rebels began manufacturing weapons in his factory. He locked up his flat and walled up the door. He hopes that the house is still standing when it someday becomes safe enough to return to Syria.

Literatur | References

Auslander, Leora; Tara Zahra. 2018. *Objects of War. The Material Culture of Conflict and Displacement.* Ithaca, NY.

Cadenbach, Christoph; Lars Reichardt; Alexander Gehring. »Schlüssel aus der Heimat«. *Süddeutsche Zeitung Magazin*, 06. Mai 2018.

https://sz-magazin.sueddeutsche.de/leben-und-gesellschaft/schluessel-aus-der-heimat-85653

Povrzanović-Frykman, Maja. 2016. Conceptualising Continuity: A Material Culture Perspective on Transnational Social Fields. *Ethnologia Fennica* 43: 45.

Schiocchet, Leonardo. 2014. Palestinian Refugees in Lebanon: Is the Camp a Space of Exception. *Mashriq & Mahjar: Journal of Middle East and North African Migration Studies* 2 (1): 130–160.

Anmerkungen

1 Matthäus 16,19: »Ich werde dir die Schlüssel des Reiches der Himmel geben; und was irgend du auf der Erde binden wirst, wird in den Himmeln gebunden sein, und was irgend du auf der Erde lösen wirst, wird in den Himmeln gelöst sein.«

Notes

1 Matthew 16:19, King James Version: "And I will give unto thee the keys of the kingdom of heaven: and whatsoever thou shalt bind on earth shall be bound in heaven: and whatsoever thou shalt loose on earth shall be loosed in heaven."

Bücher auf dem Weg nach Hause

Books Heading Home

Özlem Savaş

Abb. 1: Büchersammlung zuhause, 2021. ©Ute Langkafel

Image 1: Collecting Books at Home, 2021. ©Ute Langkafel

Wann beginnt das Leben eines Buches? Wenn es zuhause oder in einem Buchladen ins Regal gestellt wird? Wenn wir es kaufen? Beginnt es, wenn wir es lesen? Wenn es veröffentlicht wurde? Oder vielleicht in dem Moment, wenn es als Idee im Kopf der Autorin erscheint? Bücher haben mehrere Leben, wie wir selbst. Leben, die mit dem Abreisen und Ankommen, dem Reisen und Innehalten, den

When does the life of a book begin? When shelved at home or in a bookshop? When we acquire it? Does it begin when we read it? When published? Or, perhaps, the moment it appears as an idea in the author's mind? Books, like humans, have multiple lives. These begin and end with departures and arrivals, journeys and pauses, continuations and interruptions. Some of these

Abb. 2: Volles Bücherregal, 2021. ©Ute Langkafel
Image 2: Shelf full of books, 2021. ©Ute Langkafel

Fortsetzungen und Unterbrechungen beginnen und enden. Einige dieser Bücher-Leben sind miteinander verknüpft, andere stehen für sich. Einige sind geplant oder werden erwartet, andere entstehen aus heiterem Himmel. Dies ist eine Exilgeschichte von Büchern oder eine Geschichte von Büchern im Exil, die sich bewegen, zurückbleiben und warten.

Meine Bücher kamen zwei Jahre nach mir in Berlin an. Als ich nach Deutschland migrierte, packte ich meine Bücher in Kisten und schickte sie zum Haus meiner Eltern. Die leere Bibliothek, die mir, kurz bevor ich durch die Tür ging, ins Auge fiel, ist mir als letzter Blick in meine Wohnung noch in lebhafter Erinnerung. Ich packte jedoch mehrere Bücher in meinen Kof-

book-lives are connected, while others are discrete; some are planned or expected, and some are born out of the blue. This is an exilic story of books, or a story of exilic books that move, stay behind, and wait.

My books arrived in Berlin 2 years after I did. When migrating to Germany, I packed my books into boxes and dispatched them to my parents' house. The empty library shelves remain a vivid memory from that moment I walked out my apartment door for the last time. I had, however, packed several books in my suitcase, the ones that I had quickly had to choose as my favourites. *Göç Temizliği* [Migration Cleaning] by Adalet Ağaoglu, *Marcovaldo* by Italo Calvino and several novels by Barış Bıçakçı were among them. Yet, finding them in my suitcase when unpacking was not as joyful as I had hoped for. In truth, I kept these books out of my sight in my new apartment, because their presence reminded me of the others' absence. They were part of a whole, a whole that was disrupted.

After I finally moved into more stable living quarters, my parents offered to ship my books. A dream-like possibility. But uneasy questions also quickly entered my mind. Am I really settling in Berlin? Will I be allowed to settle? What if I move again?

Bücher auf dem Weg nach Hause

fer, die ich in diesem Moment schnell als meine Liebsten auswählen musste. *Göç Temizliği* (Migrationsreinigung) von Adalet Ağaoglu, *Marcovaldo* von Italo Calvino und mehrere Romane von Barış Bıçakçı gehörten dazu. Doch als ich sie beim Auspacken in meinem Koffer fand, war die Freude nicht so groß, wie ich gehofft hatte. In Wahrheit habe ich diese Bücher in meiner neuen Wohnung versteckt gehalten, denn ihre Anwesenheit erinnerte mich an die Abwesenheit der anderen. Sie waren Teil eines Ganzen, eines Ganzen, das gestört war.

Nachdem ich endlich in eine Wohnung gezogen war, in der ich länger bleiben würde, boten mir meine Eltern an, meine Bücher zu mir zu schicken. Eine traumhafte Möglichkeit. Aber schnell kamen mir auch sorgenvolle Fragen in den Sinn. Werde ich in Berlin bleiben, mich hier niederlassen? Werde ich mich hier niederlassen dürfen? Was ist, wenn ich wieder umziehe? Die Möglichkeit, meine Bibliothek wiederherzustellen, überschnitt sich mit der Möglichkeit, zu begreifen, zu akzeptieren und laut zu sagen: »Ich lebe hier.« Nach ein wenig Angst und Verwirrung war ich von beiden Möglichkeiten angetan. Mein Vater öffnete die Kisten mit den Büchern, las mir jeden Titel vor, und ich sagte in einem stundenlangen WhatsApp-Gespräch ja oder nein. Die ausgewählten Bücher wurden wieder eingepackt und nach Berlin verschickt. Unser Wiedersehen fand auf dem Zollamt statt, das sowohl die Grenze als auch die Kon-

The possibility of restoring my library overlapped with the possibility of apprehending, accepting, and stating aloud "I live here." After feeling a bit anxious and confused, both possibilities started to appear tempting. My father opened the boxes of books, read each title to me, and I said yes or no, in hours of WhatsApp conversations. The chosen books were repacked and shipped to Berlin. Our reunion took place at the customs office, which was both the border and the zone of contact between my books and me. Ripping the boxes open, I met my books again after 2 years, under the eyes of the customs' personnel. I rescued my books from this alienating border

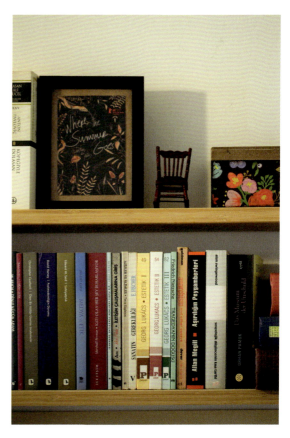

Abb. 3: Die Bücher in ihrem neuen Zuhause, 2021.
©Ute Langkafel
Image 3: The books in their new home, 2021.
©Ute Langkafel

taktzone zwischen mir und meinen Büchern war. Ich riss die Kisten auf und sah meine Bücher nach zwei Jahren wieder, unter den Augen der Zollbeamt:innen. Ich rettete meine Bücher so schnell wie möglich von dieser entfremdenden Grenze und brachte sie nach Hause. Ich verteilte die Bücherstapel in meiner 50 m² großen Wohnung, betrachtete und roch sie einen ganzen Tag lang. Schließlich wischte ich den Staub von den leeren Regalen und ordnete meine Bücher mit Gefühlen von Freude und Trauer, Müdigkeit und Aufregung, Gewohnheit und Neuheit. Meine Bibliothek richtete sich ein weiteres Mal ein. Während ich mein Gedächtnis zwinge, diese Geschichte zu erzählen, frage ich mich immer wieder: Warum ist das Vorhandensein oder die Abwesenheit von Büchern so wichtig inmitten der wichtigeren Themen des täglichen Lebens, der Arbeit und der Politik? Welche Beziehung haben Exilant:innen zu ihrer persönlichen Büchersammlung?

Die Geschichte meiner Bücher ist nur eine von Tausenden von Büchern, die sich im Zuge der derzeitigen Migration aus der Türkei bewegen, zurückbleiben und warten. Die eskalierende politische Unterdrückung und die Unruhen, vor allem nach der Gezi-Bewegung 2013 und dem Putschversuch 2016, haben viele Menschen dazu veranlasst, das Land zu verlassen, insbesondere Akademiker:innen, Künstler:innen, Journalist:innen und Studierende, die mit Büchern gearbeitet, studiert und gelebt haben. Der Weggang hat die bittere Aufgabe mit sich gebracht, die Bücher in Kisten zu verpacken und sie einem unvorhersehbaren Schicksal zu überlassen. Die Umsiedlung ging mit einer Reihe von Gefühlen und Praktiken im Umgang mit den persönlichen

as quickly as possible and brought them home. I spread the piles of books all over my 50 m² apartment, looked at them and smelled them for an entire day. Finally, I cleaned the dust off the empty shelves and ordered my books, in changing states of joy and grief, tiredness and excitement, out of habit and with a sense of novelty. My library settled in one more time. While forcing my memory to tell this story, I keep asking myself: Why is the presence or the absence of the books so important amid more urgent issues of daily life, work and politics? What is the relationship of an exile to their personal collection of books?

The story of my books is only one of thousands of books that move, stay behind and wait during the recent migration out of Turkey. The escalating political oppression and turmoil, largely after the 2013 Gezi movement, and the attempted coup in 2016 prompted many people to leave the country, especially academics, artists, journalists and students, who work, study and live with books. Leaving has necessitated the bitter task of packing the books into boxes and abandoning them to an unforeseeable destiny. Resettlement has corresponded with a range of feelings and practices related to the personal collections of books: mourning for the books that stayed behind, shipping the boxes of books, waiting to reunite with the books, relocating them in new libraries, starting to collect books again and establishing online groups for swapping printed books. Amid the profound experiences of loss and uncertainty, it has been commonly voiced by intellectuals exiled from Turkey that

Büchersammlungen einher: Trauer um die zurückgebliebenen Bücher, Versand der Bücherkisten, Warten auf das Wiedersehen mit den Büchern, Umplatzierung der Bücher in den neuen Bibliotheken, Wiederaufnahme des Sammelns von Büchern und Einrichtung von Online-Gruppen zum Austausch von Büchern. Inmitten der tiefgreifenden Erfahrungen von Verlust und Ungewissheit haben Intellektuelle, die aus der Türkei ins Exil gegangen sind, häufig geäußert, dass das »Zuhause dort ist, wo Deine Bücher sind«. Ihr Wunsch, persönliche Büchersammlungen wiederherzustellen, entspringt weniger den Notwendigkeiten von Arbeit und Studium, sondern dem Wunsch, weiterhin mit den Büchern zusammenzuleben, die sich im Laufe eines Lebens angesammelt haben. Ich kann nicht vergessen, was ein Freund von mir, der vor Kurzem die Türkei verlassen hat, zu mir sagte, nachdem er mit einem langfristigen Vertrag in eine Mietwohnung in Berlin gezogen war: »Das ist das erste Mal seit Jahren, dass meine Bücher und ich wieder am selben Ort leben können.«

Unser Wunsch, unsere Bibliotheken wiederherzustellen, bezieht sich nicht nur auf unseren Wunsch nach einem Zuhause, sondern auch auf die Bücher, die wir in Kisten zurückgelassen haben. Mit anderen Worten: Im Exil sind nicht nur die Menschen, sondern auch ihre Bücher, die sie in den Kisten zurückgelassen haben, während sie in einem Schwebezustand auf ein neues Zuhause warten. Eine Freundin und Kollegin von mir, die es geschafft hat, ihre Bücher nach Berlin zu bringen, erinnert sich traurig an den Tag, an dem sie die Kisten mit den Büchern öffnete, die ihre Schwester bei ihrer Migration vor langer Zeit zurückgelassen hatte. Mit dem Bild der beschädigten und ver-

"home is where your books are." This desire to restore personal book collections arises less from the necessities of work and study than from the wish to continue living together with books that have accumulated over a lifetime. I cannot forget what a friend of mine, who recently left Turkey, said to me after obtaining a longer-term rental apartment in Berlin: "This is the first time for years that my books and I can live in the same place again."

Our desire to restore our libraries pertains to making a home not only for ourselves but also for the books we abandoned in boxes. Put differently, not only people live in exile but also their books, which stay behind, stuck in boxes, waiting in limbo for a home that was yet to come. A friend and colleague of mine, who arranged to have her books transported to Berlin, sadly remembers the day she opened the boxes of books her sister had left behind when she migrated a long time ago. Thinking of the image of the damaged and mouldy books left in in boxes for about 10 years, she says: "I did not want the same thing to happen to my books." To avoid something similar happening to her, another friend and colleague, who had to leave her books in Turkey, gave her books to a friend, after waiting and hoping to bring them to Berlin for a whole 2 years. With her books now resettled in her friend's library, she says: "My feelings of grief about my books have lessened, but have not gone away completely."

Adalet Ağaoğlu wrote her autobiographical novel *Göç Temizliği* [Migration Cleaning] about the day she spent clearing out her workroom, before migrating

schimmelten Bücher im Kopf, die etwa zehn Jahre lang in den Kisten warteten, sagt sie: »Ich wollte nicht, dass mit meinen Büchern dasselbe passiert.« Aus demselben Grund hatte eine andere Freundin und Kollegin, die ihre Bücher in der Türkei zurücklassen musste, ihre Bücher einer Freundin geschenkt, nachdem sie zwei Jahre lang darauf gewartet und gehofft hatte, sie nach Berlin bringen zu können. Ihre Bücher sind nun in der Bibliothek ihrer Freundin untergebracht, und sie erzählte: »Meine Trauer um meine Bücher ist weniger geworden, obwohl sie nicht ganz verschwunden ist.«

Adalet Ağaoğlu schreibt in ihrem Memoiren-Roman *Göç Temizliği* über den Tag, an dem sie ihr Arbeitszimmer »aufräumte«, bevor sie 1983 von Ankara nach Istanbul migrierte. Sie musste alle Bücher, Notizbücher, Zeitschriften, Briefe und Fotos durchsehen, die seit zwanzig Jahren in den Regalen, Schränken und Schubladen lagen und sie musste diejenigen auswählen und einpacken, die sie mitnehmen wollte, und den Rest wegwerfen. Sie musste versuchen, die Erinnerungen aufzuarbeiten, die diese Stapel von Papieren enthüllten. Einerseits beschreibt sie diese Aufgabe, die sie bis zum letzten Tag vor dem Eintreffen der Umzugsfirma aufgeschoben hatte, als seelisch erleichternd: »Es wäre dumm, nicht so leicht wie möglich zu reisen« (30). Andererseits fiel es ihr schwer, die Sammlung, die sich im Laufe ihres Lebens angesammelt hatte, zu entrümpeln. Sie glaubt, dass »wir das verborgene Gesicht von morgen nicht sehen können, ohne eine Auswahl von gestern zu treffen« (27). Gleichzeitig fürchtet sie das Vergessen. Und doch muss es getan werden: »Wie ein unausweichlicher Sprung ins Wasser inmitten von Eis und Sturm, schließe ich meine Augen und tauche ein in die Stapel

from Ankara to Istanbul in 1983. She had to go through all her books, notebooks, magazines, letters and photos that had inhabited the shelves, cabinets, and drawers for the previous 20 years. She had to choose and pack the ones that she would take with her and throw away the rest. And she had to attempt to catch up with the memories these piles of papers triggered. On the one hand, she described this task, which she put off until the last day before the removal company came, as an unburdening of the soul: "It would be foolish not to go as lightly as possible" (30). On the other, she felt heavy hearted at decluttering the collection that had accumulated during her lifetime. She believes that "we cannot see the hidden face of tomorrow without making a selection about yesterday" (27), but, at the same time, she fears forgetting. Yet, it has to be done: "Like inescapably diving into water in the midst of ice and storm, I close my eyes and dive into the piles of books, papers, files and notebooks in my room" (34).

Especially when they move do people realise how heavy books are. Books are among those things whose materiality is apparently evident but also confusing. Is it the combination of written words or the bundled papers that contain words, what we call a book? The question has become more interesting with the emergence and spread of e-books. I suggest that the perceived materiality of books varies depending on the type of interactions we have with them. While reading and contemplating the flow of words, the physical book can seem to disappear. The bodily, tangible

Abb. 4: Notizbücher, 2021. ©Ute Langkafel

Image 4: Notebooks, 2021. ©Ute Langkafel

von Büchern, Papieren, Akten, Notizbüchern in meinem Zimmer« (34).

Wie schwer Bücher sind, wird beim Umziehen spürbar. Bücher gehören zu den Dingen, deren Materialität offensichtlich aber verwirrend ist. Ist es die Kombination aus geschriebenen Wörtern oder die gebündelten Papiere, die Wörter enthalten, die wir ein Buch nennen? Die Frage ist mit dem Aufkommen und der Verbreitung von digitalisierten Büchern noch interessanter geworden. Ich denke, dass die wahrgenommene Materialität von Büchern je nach der Art unserer Interaktion mit ihnen variiert. Während des Lesens und des Nachdenkens über den Fluss der Worte kann das physische Buch verschwinden. Die körperlichen, greifbaren Aspekte des Lesens können durch die gewohnheitsmäßige Koor-

aspects of reading can become imperceptible through the habitual coordination of papers, eyes and hands. Even while sharing the home with them, the materiality of shelved books might go unnoticed. A well-organised library might turn into an overly familiar and thus overlooked image that covers the walls. The materiality of books becomes prominent, however, when they are acquired and then abandoned, packed and unpacked. The physical space that books occupy becomes enormous when they are taken out of the shelves. The accumulated weight of books is felt most immediately when they are packed in boxes and then lifted. During moving, the piles of books become a burden, both physically and emotionally. Why then do

dination von Papier, Augen und Händen unmerklich werden. Sogar wenn mit ihnen das Haus geteilt wird, kann die Materialität der Bücher im Regal unbemerkt bleiben. Eine gut organisierte Bibliothek kann zu einem allzu vertrauten und daher übersehenen Bild werden, das die Wände bedeckt. Die Materialität von Büchern tritt jedoch in den Vordergrund, wenn wir sie erwerben und wegwerfen, ein- und auspacken. Der physische Raum, den Bücher einnehmen, wird riesig, wenn sie aus den Regalen genommen werden. Das angehäufte Gewicht von Büchern wird wahrgenommen, wenn sie, in Kisten verpackt, angehoben werden. Bei einem Umzug werden die Bücherstapel zu einer physischen und emotionalen Belastung. Warum machen wir uns dann die

we even trouble ourselves to move our books or to mourn those that stay behind? Why do we collect books in the first place?

As collectors can vouchsafe, an appreciation of books stretches far beyond the benefits and pleasures of reading. Acquired from different places and at different times, organised and reorganised into a library on many occasions, a collection of books represents a constellation of life-fragments. Books are intimate and silent witnesses to life. In "Unpacking My Library", Walter Benjamin explores the passions of a book collector, by inviting the reader to join him "among piles of volumes that are seeing daylight again after 2 years of darkness" (59), on the day he reunites

Abb. 5: Walter Benjamin zwischen Herta Müller und Benedict Anderson, 2021. ©Ute Langkafel

Image 5: Walter Benjamin between Herta Müller and Benedict Anderson, 2021. ©Ute Langkafel

Mühe, mit unseren Büchern umzuziehen, oder warum trauern wir um die, die zurückbleiben? Warum sammeln wir überhaupt Bücher?

Sammler sind sich einig, dass die Wertschätzung von Büchern nicht auf den Nutzen und das Vergnügen des Lesens beschränkt ist. Eine Büchersammlung ist eine Konstellation von Lebensfragmenten, die an verschiedenen Orten und zu verschiedenen Zeiten erworben, organisiert und zu einer Bibliothek umorganisiert wurden. Bücher sind intime und stille Zeugen des Lebens. In »Ich packe meine Bibliothek aus« erkundet Walter Benjamin die Leidenschaften eines Büchersammlers, indem er die Leser:innen einlädt, ihn zu begleiten und sich zwischen die »Stapel [...] nach zweijähriger Dunkelheit wieder ans Tageslicht beförderter Bände [...] zu versetzen [...]« (388), an dem Tag, an dem er mit seiner Büchersammlung in seiner neuen Wohnung wiedervereinigt wird. Benjamins Leidenschaft, »grenzt [...] ans Chaos [...] der Erinnerungen« (388), an die gewohnte »Regellosigkeit einer Bibliothek« (389). Die Bücher, die sich um ihn herum stapeln, sind noch nicht von der »leise[n] Langeweile der Ordnung« (388) berührt, noch nicht ins Regal gestellt und offenbaren so die tatsächliche Unordnung, die seine Sammlung zusammenhält: »Zufall, Schicksal, die das Vergangene vor meinem Blick durchfärben, sie sind zugleich in dem gewohnten Durcheinander dieser Bücher sinnenfällig da. Denn was ist dieser Besitz anderes als eine Unordnung, in der Gewohnheit sich so heimisch machte, daß sie als Ordnung erscheinen kann?« (388)

Die gelebte Materialität von Büchern ist die einer Kontinuität, die durch Fragmente, Unterbrechungen und Brüche erschaffen

with his book collection in his new apartment. Benjamin's passion "borders on the chaos of memories" (60), and on the habitual "confusion of a library" (60). The books that pile around him, "touched by the mild boredom of order" (59) but not yet shelved, disclose the actual disorder that assemble his collection: "[T]he chance, the fate, that suffuse the past before my eyes are conspicuously present in the accustomed confusion of these books. For what else is this collection but a disorder to which habit has accommodated itself to such an extent that it can appear as order?" (60)

The lived materiality of books is one of a continuity created through fragments, interruptions and ruptures. Yet, no matter how imperfect, accidental or inattentive they can be, all libraries are organised in some way or another, just like our memories are often organised into linear and propositional life histories. The habitual everyday order of both a library and an ordinary life masks how fragmented and scattered both things actually are. This illusion of an uninterrupted continuity lasts only until an unexpected turn of events disrupts the ordinary, because it is the ordinary that stitches the parts and pieces together. The interruption of habit breaks down the accustomed constellation of both books and life-fragments. For the exile, what is disrupted is this habitual everyday order, life as it is, no matter how imperfect and accidental it might be. For the exile, the ordinary present is so evident and palpable that every condition, occurrence, and feeling in it becomes noticeable and demands attention.

Abb. 6: Jede Buchsammlung hat ihre eigene Ordnung, 2021. ©Ute Langkafel

Image 6: Every book collection has its own sense of order, 2021. ©Ute Langkafel

wird. Doch egal wie unvollkommen, zufällig oder gleichgültig sie auch erscheinen mögen, alle Bibliotheken sind auf die eine oder andere Weise organisiert, so wie unsere Erinnerungen oft in linearen und in sich logischen Lebensgeschichten organisiert sind. Die gewohnte alltägliche Ordnung einer Bibliothek und eines gewöhnlichen Lebens täuscht darüber hinweg, wie fragmentiert und verstreut beide in Wirklichkeit sind. Diese Illusion einer ununterbrechenen Kontinuität hält nur so lange an, bis eine unerwartete Wendung das Gewohnte zerbrechen lässt, denn es ist das Gewohnte, das Teile und Stücke zusammenfügt. Durch die Unterbrechung der Gewohnheit bricht die vertraute Konstellation von Büchern und Lebensfragmenten zusammen. Was für im Exil Lebende gestört ist, ist die

As a migration scholar in exile, I am puzzled by how many discourses – academic or popular – regard practices of remembering, reinterpreting and recollecting the past as symptoms of an exilic fixation on the past, making the migrant and the exile the most famous subjects of melancholia and nostalgia. I am puzzled because the present is most evident, palpable and urgent to the exile, more than for anyone else. Once its habitual ordinariness is disrupted and unmasked, the present reveals itself in its entirety of confusion and immediacy, demanding overly self-reflexive attention and commitment. If there is any timeframe in which the exile is stuck, it is not the past but the present state of constantly working on and

Bücher auf dem Weg nach Hause

gewohnte Alltagsordnung, das Leben, wie es ist, egal, wie unvollkommen und zufällig es auch sein mag. Für im Exil Lebende ist die gewöhnliche Gegenwart so offensichtlich und greifbar, dass jeder Zustand, jedes Tun und jedes Gefühl spürbar wird und Aufmerksamkeit erfordert.

Als Migrationswissenschaftlerin im Exil stelle ich mit Verwunderung fest, wie viele Diskurse – akademische wie öffentliche – Praktiken des Erinnerns, der Neuinterpretation und der Rückbesinnung auf die Vergangenheit als Symptome einer exilischen Fixierung auf die Vergangenheit betrachten und Migrant:innen und Exilant:innen zu den berühmtesten Subjekten von Melancholie und Nostalgie machen. Ich bin verwirrt, weil die Gegenwart für im Exil Lebende die offensichtlichste, greifbarste und dringendste Zeit ist, mehr als für jeden anderen. Sobald ihre gewohnte Alltäglichkeit unterbrochen und entlarvt ist, offenbart sich die Gegenwart in ihrer ganzen Verwirrung und Unmittelbarkeit und fordert übermäßiges Engagement und selbstreflexive Aufmerksamkeit. Daher interpretiere ich unsere Bemühungen um die Wiederherstellung unserer Bibliotheken, um die Erinnerung und Neuordnung unserer verstreuten Lebensfragmente nicht als melancholische Bindung an die Vergangenheit. Dass wir unsere Bücher nach Hause bringen, ist vielmehr eine materielle und emotionale Investition in die Gegenwart und die Zukunft. Das Gefühl der Kontinuität, das die Büchersammlung vermittelt, ist eine hoffnungsvolle Möglichkeit, hier und jetzt ein Zuhause zu schaffen.

Das Packen und Auspacken einer Bibliothek ist ein Packen und Auspacken von Lebensfragmenten. Die gelebte Materialität

waiting for the future. Thus, I do not interpret our efforts in restoring our libraries as well as recollecting and reorganising our scattered life-fragments as a melancholic attachment to the past. Rather, bringing our books home is a material and emotional investment in the present and the future. The sense of continuity a collection of books gives is a hopeful possibility for making a home, here and now.

Packing and unpacking a library means packing and unpacking life-fragments. The lived materiality of books is one of continuity, which is not subsumed under but resists a linear, propositional life narrative. A continuity with fragments, interruptions, ruptures, with beginnings and endings. A continuity in an exilic life. The story of exilic books, whether brought to a new home or left behind, whether relocated in a new library, waiting in boxes, or en route, is the story of exiles on their way home.

von Büchern ist von Kontinuität geprägt. Eine Kontinuität, die nicht unter eine lineare, in sich logische Lebenserzählung geordnet werden kann, sondern sich dieser widersetzt. Eine Kontinuität mit Fragmenten, Unterbrechungen, Brüchen, mit Anfängen und Enden. Eine Kontinuität in einem exilischen Leben. Die Geschichte der Bücher im Exil, ob sie nun in ein neues Zuhause gebracht oder zurückgelassen wurden, ob sie in einer neuen Bibliothek untergebracht wurden, in Kisten warten oder unterwegs sind, ist die Geschichte der im Exil Lebenden, die auf dem Weg zu einem Zuhause sind.

Literatur | References

Ağaoğlu, Adalet. 1985. *Göç Temizliği* (Migration Cleaning). Istanbul.

Benjamin, Walter. 1968. Unpacking My Library. In: Hannah Arendt: *Illuminations*. Translated by Harry Zohn, 59–67. New York.

Benjamin, Walter. 1972. Ich packe meine Bibliothek aus. Eine Rede über das Sammeln. In: Tillman Rexroth: *Walter Benjamin. Gesammelte Schriften IV. Teil 1*, 388–396. Frankfurt a.M.

Vom Koffer zum Fluchtgepäck und wieder zurück

From Suitcase to Flight Baggage and Back

Veronika Reidinger | Anne Unterwurzacher

Koffer und sonstige Gepäckstücke sind unsere ständigen Begleiter, wenn wir unterwegs sind. Wer beruflich verreist, andernorts Freunde oder Verwandte besucht oder in den Urlaub fährt, verwendet Gepäck in den unterschiedlichsten Formen und Gestalten. Wir packen nicht nur für längere Abwesenheiten ein, auch alltäglich befüllen wir Hand- oder Schultaschen mit den als notwendig erachteten Dingen. Koffer, Taschen und sonstige Gepäckstücke dienen uns dabei als Behälter für Dinge, die wir an andere Orte transportieren möchten. Mit dem Massentourismus, der im 20. Jahrhundert nach und nach alle sozialen Schichten erfasste, ist das Einpacken zu einer vielfach erprobten sozialen Praktik geworden.

Doch wie verhält es sich, wenn Menschen flüchten müssen? Dieser Frage möchten wir im Folgenden nachgehen. Wir lassen dabei die Erzählungen von geflüchteten Personen einfließen, die wir im Rahmen des Projektes »Mobile Dinge« zu ihrer Flucht und den von ihnen mitgebrachten oder zurückgelassenen Dingen befragt haben.[1] Ergänzend beziehen wir uns auf zwei Ausstellungen: die Sonder-

Suitcases and other pieces of luggage are constant companions when we are on the move. Anyone who travels on business, visits friends or relatives or goes on a long-awaited vacation has used suitcases in a wide variety of shapes and forms. We do not just pack for longer absences; we also fill handbags or school bags with things we consider necessary to make it through the day. Suitcases, bags and other pieces of luggage serve as containers for things we want to transport to other places. With the advent of mass tourism, which has spread gradually to affect all social classes, packing a suitcase has become a tried and tested social practice. But what happens when people have to flee? We would like to pursue this question here, incorporating the stories of refugees whom we asked about their flight and the things they brought with them or left behind, as part of the Austria-based project *Mobile Dinge* [Mobile Things].[1] This narrative is augmented by references to two exhibitions, *Little Vienna in Shanghai*, which showed at the Jewish

ausstellung *Die Wiener in China. Fluchtpunkt Shanghai* im Jüdischen Museum Wien² und die Dauerausstellung *Die Küsten Österreichs* des Volkskundemuseums Wien.³

Museum Vienna,² and the current permanent exhibition *The Shores of Austria* in the Vienna *Volkskunde-Museum* [Ethnographic Museum].³

»Irgendwie habe ich auch Abschied genommen« – Vom Einpacken und Auswählen

"Somehow I also said goodbye" – On Packing and Choosing

Der Koffer bzw. das Gepäckstück kann als Container gesehen werden, der die eingepackten Dinge nicht nur einschließt, sondern zugleich all das ausschließt, was nicht mehr hineinpasst (vgl. Baur 2009: 340ff.). Damit wird die Handlung des Einpackens relevant. Beim Einpacken wählen wir die Dinge anhand dessen aus, was unserer Vorstellung nach zukünftig gebraucht wird. Die Selektion der Dinge wird somit zu einer »Übung in Antizipation« (Harlan 2018: 66), in der wir uns eine manchmal mehr, manchmal weniger ungewisse Zukunft vorstellen. Im Kontext von Flucht stellt sich die Frage des Einpackens und der Dinge, die mitgenommen werden sollen, in besonderer Weise. Das Packen wird strategisch nicht nur auf eine unbestimmbare Zukunft ausgerichtet, sondern die zurückgelassenen Dinge sind eventuell zu einem späteren Zeitpunkt nicht mehr verfügbar. In unserem Projekt »Mobile Dinge« erzählte etwa Reza⁴ vom »Abschied nehmen« von (Spiel-)Sachen während des Einpackens für die Flucht aus dem Iran aus seiner damaligen kindlichen Perspektive:

> … irgendwie habe ich auch Abschied genommen. … ich habe sortiert. Okay, die brauche ich später, das weiß ich, wenn ich älter bin, werde ich nicht mehr damit

A suitcase or a piece of luggage can be seen as a container that not only encloses the packed things but also excludes what does not fit inside (cf. Baur 2009: 340ff.). This makes the question of packing relevant. When packing, we choose things based on what we think will be needed in the future. The process of packing thus becomes an "exercise in anticipation" (Harlan 2018: 66), in which we imagine a sometimes more, sometimes less uncertain future. In the context of flight, the question of packing and the things one wants to take along acquires an additional dimension. Not only is packing oriented strategically towards an indefinite future; potentially, the things left behind will be lost indelibly for that future. In the project *Mobile Things*, for example, Reza⁴ describes "saying goodbye" while packing as a child, before fleeing from Iran:

> Somehow I also said goodbye … I sorted my possessions. Okay, I'll not need that later, I know; when I'm older I won't play with that anymore. I will sell you or give you away. But I'll put you here, we'll see each other again. (Reza S.)

spielen. Dich werde ich verkaufen oder verschenken. Aber du kommst hier hin, wir sehen uns wieder. (Reza S.)

Die Erzählung liefert einen intimen Einblick in den Akt des Einpackens als personalisierte Aushandlung, in der Reza seiner kindlichen Perspektive Vorstellungen von zukünftigen erwachsenen Bedürfnissen gegenübergestellt und dementsprechend Dinge ein- und aussortiert.

Menschen, die aus den unterschiedlichsten Gründen flüchten müssen, können sich jedoch nicht immer überlegen, was sie mitnehmen möchten. Finden sie noch die Zeit, wichtige Vorbereitungen zu treffen oder müssen sie plötzlich und unvorbereitet einen Ort verlassen? Sehen sie sich gezwungen, ihr Hab und Gut vor Ort zurückzulassen? Tragen sie beim Packen die Idee des Zurückkommens in sich, können sie bestimmte Dinge bei Bekannten oder Nachbarn (zwischen-)lagern? Es macht für das Einpacken einen Unterschied, ob die Flucht als temporäres oder dauerhaftes Abschiednehmen gefasst wird, wie auch folgendes Zitat verdeutlicht:

> Einige von ihnen wussten, wohin sie gingen, aber die meisten wussten nur, dass sie »raus« gingen, wo auch immer das war, an »einen sicheren Ort«. Sie ließen einen Monatsvorrat an Katzenfutter für ihre Haustiere auf dem Boden liegen, ihre Wäsche im Hof hängend, ihr Schmuck unter einem Holzzaun vergraben, ihre Wintervorräte sorgfältig im Keller eingelagert; sie alle wussten, dass sie zurückkommen würden, »bald«. (Mertus et al 1997:22)

The narrative provides an intimate insight into the act of packing as a personal negotiation, in which Reza grapples with future and potential adult needs from a child's perspective and sorts his possessions accordingly.

That said, people who have to flee cannot always reflect calmly on what to take with them. Do they find the time to make important preparations or do they have to leave suddenly and unprepared? Are they being forced to leave their belongings behind? Does the process of packing include the idea of returning? Can certain personal effects be placed in the (temporary) care of friends or neighbours? It thus makes a difference when packing whether flight is seen as temporary or permanent, as the following statement illustrates:

> Some of them knew where they were headed, but most knew only that they were going "out", wherever that was, to "some safe place". They left a month's supply of cat food on the floor for their pets, their laundry hanging in the yard, their jewellery buried underneath a wooden fence, their winter preserves carefully stored in the cellar; they all knew that they would be back "soon." (Mertus et al. 1997: 22)

This passage is cited from a book titled *The Suitcase*, which describes people preparing to flee from Bosnia and Croatia at the start of the Yugoslav Wars, which began in the early 1990s. The protagonists made these arrangements with the intention of returning. Ideas of returning, or, on the

Dieses Zitat ist dem Buch *The Suitcase* (Der Koffer) entnommen und beschreibt Vorbereitungshandlungen zur Flucht aus Bosnien und Kroatien in Folge der dortigen Kriege in den 1990er Jahren, die auf ein Zurückkommen angelegt waren. Vorstellungen einer Rückkehr oder die eines endgültigen Abschieds üben somit einen Einfluss darauf aus, was mobil gemacht und was (vorübergehend) zurückgelassen wird. Während im vorangestellten Zitat etwa die Monatsration an Katzenfutter zurückgelassen wurde, wurde in einer anderen verschriftlichten Erzählung, die wir im Projekt »Mobile Dinge« gesammelt haben, das Haustier selbst zum Teil des Fluchtgepäcks:

contrary, of a final farewell thus influence what is taken and what is left behind – at least for the time being. While the preceding citation lists the monthly cat-food ration as something that was not taken, in another written narrative collected in the "Mobile Things" project, the pet itself became part of the baggage of flight:

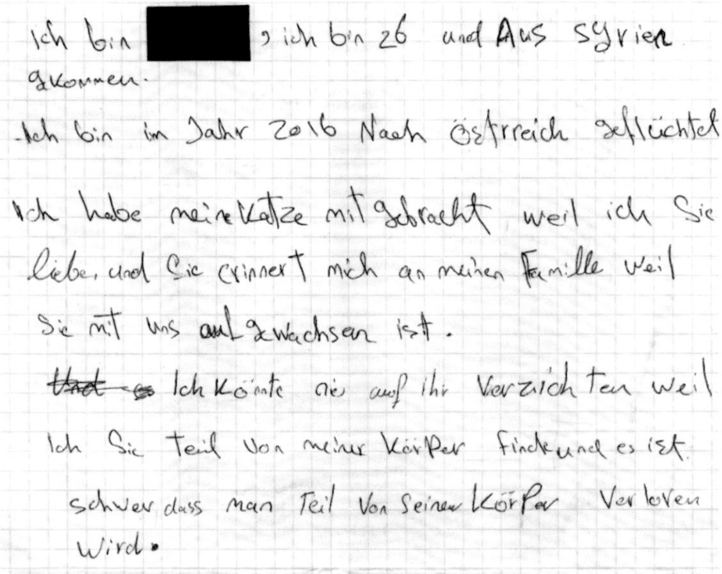

Abb. 1: Schriftliche Erzählung zu Dingen auf der Flucht, 2019: Projekt »(Nicht) im Gepäck?«, Ilse Arlt Institut für Soziale Inklusionsforschung

Image 1: Written note on things in the process of flight. From the 2019 project "(Nicht) im Gepäck?" [(Not) in the bag?], ©Ilse Arlt Institute for Social Inclusion Research

I am [Name redacted], I'm 26, and I come from Syria.
I fled to Austria in 2016.
I took my cat with me because I love her, and she reminds me of my family, because she grew up with us.
~~And it~~ I could never do without her because for me she is like a part of my body, and it is difficult if a part of one's body gets lost.

Vom Koffer zum Fluchtgepäck

Die Erzählung verdeutlicht, dass eine Katze (oder auch Dinge) Teil der eigenen Person werden können. Sie nicht mitzunehmen, wird damit undenkbar oder zumindest sehr schmerzhaft. Abseits von Überlegungen, ob bestimmte Dinge ›nützlich‹ sein oder einen ›Gebrauchswert‹ in einer unbestimmten Zukunft haben werden, spielen somit auch emotionale Verbindungen zu den unterschiedlichen Dingen beim Prozess des Auswählens eine Rolle.

The narrative also demonstrates how much a cat, or indeed other things, can become part of oneself. Not taking them along becomes unthinkable or at least very painful. Alongside considering whether particular things might be 'useful' or have any 'utility value' in an indefinite future, emotional connections to such diverse things also play a role in the selection process.

»Ich war nur mit einem Rucksack unterwegs« – Von Mobilitätsmöglichkeiten

"I was only carrying a backpack" – On Mobility Options

Das Reisegepäck ist eng an die Art (und Möglichkeit) des Reisens selbst geknüpft. Umgekehrt prägt das Gepäck, das wir mit uns führen, die Art und Weise, wie wir uns fortbewegen können: »Der Koffer ist nicht nur ein Behältnis, sondern ein Generator, der dem Handeln des Reisenden Form und Richtung gibt« (Löfgren 2016: 134). Historisch betrachtet bezog sich der Begriff »Koffer« bis vor rund 150 Jahren auf Kisten- und Truhenmöbel, die als die Vorläufer der modernen Koffer gelten (Selheim 2019: 12). Die Gestalt des Reisegepäcks hat sich mit dem Aufkommen neuer Transportmittel wesentlich gewandelt: Kutschenkoffer wurden außen auf dem Dach bzw. seitlich oder hinten auf eigens dafür vorgesehenen Gepäckvorrichtungen mitgeführt. Sie mussten wetterfest sein und hatten für diesen Zweck einen gewölbten Deckel und einen regenabweisenden Farb- und Lackanstrich (Mihm 2001: 51). Mit den aufkommenden Bahnreisen konnten Menschen dann große Behältnisse ohne Gewichtsbegrenzun-

Luggage is closely linked to the modes and the possibilities of travel itself. Conversely, the luggage we carry with us shapes how we get around: "The suitcase is not just a container but a generator, providing shape and direction for the traveller's actions" (Löfgren 2016: 134). Viewed historically, the term 'suitcase' was used until around 150 years ago to mean chests and chest-like furniture, which only gradually came to look like the forerunners of modern suitcases (Selheim 2019: 12). The shape of luggage has changed significantly with the advent of new means of transport: Chests were strapped to the outside or to the roof of carriages by means of constructions especially designed for this purpose. They had to be weatherproof and had domed lids, rain-repellent paint and varnish coatings for this purpose (Mihm 2001: 51). With the advent of rail travel, large containers were transported in separate luggage cars with no weight restrictions. Sturdy wood-

gen in separaten Gepäckwagen transportieren lassen. Es kamen damals lange, robuste Koffertruhen aus Holz mit flachem Deckel in Mode, die stapelbar waren. Für das Bahnabteil hingegen sollte das verwendete Gepäck handlicher sein, damit Reisende es selbst tragen bzw. heben und im Gepäcknetz oder unter dem Sitz verstauen konnten. Auf diese Weise entstand der Handkoffer (ebd.: 55). Die heute global verbreiteten Reise-Trolleys wiederum wurden für die Reise im Flugzeug konzipiert. Während diese im unwegsamen Gelände zum Hindernis werden, ermöglicht ein Rucksack mehr Bewegungsfreiheit und Flexibilität und ist daher der ideale Begleiter für das Unterwegs-Sein zu Fuß.

In der Schausammlung *Die Küsten Österreichs* illustrieren die beteiligten Kurator:innen

en trunks with flat lids that were stackable became fashionable. Hand luggage, on the other hand, had to be more manageable so that travellers could carry it themselves and stow it in the provided luggage net or under the seat. This is how the modern suitcase came about (Mihm 2001: 55). The modern travel trolleys found today around the world are, in turn, designed for travel by airplane. That said, these become a hindrance in rough terrain, while a backpack allows for more freedom of movement and flexibility and is therefore the ideal companion when on foot.

In the display room entitled "Wohnkultur" ['Home/Housing Culture'] in *The Shores of Austria* exhibition, the curators illustrated the change in mobile contain-

Abb. 2: Themenraum »Wohnkultur« der Schausammlung *Die Küsten Österreichs* im Volkskundemuseum Wien, 2021. ©Patrizia Gapp

Image 2: A themed-room on the culture of homes and housing in the permanent exhibition *The Shores of Austria*, Vienna Volkskunde-Museum, photograph, 2021. ©Patrizia Gapp

den Wandel der mobilen Dingcontainer, indem sie den im Themenraum »Wohnkultur« ausgestellten Truhen eine auf der Flucht verwendete Reisetasche und eine Plastiktüte gegenüberstellen. Truhen zählten zur beliebten Ausstattung in adeligen, bürgerlichen und bäuerlichen Wohnräumen. Es wurden darin alle Arten von Gegenständen aufbewahrt, häufig dienten sie auch als Sitzgelegenheit und erfüllten somit unterschiedliche Funktionen.[5] Truhen begleiteten ihre Besitzer:innen an andere Orte. Sie enthielten beispielsweise die Aussteuer, wenn eine Frau nach der Heirat in einen neuen Haushalt zog. Gesindetruhen beherbergten zumeist all das (wenige), das Knechte und Mägde besaßen (Mihm 2001: 18). Die ausgestellte Reisetasche und die Plastiktüte hingegen umhüllen nur das, was vom Besitz der Geflüchteten übriggeblieben ist.

Während für das Bewegen der schweren Truhen eine entsprechende Infrastruktur benötigt wurde, sind Reisetasche und Plastiktüte einfacher zu mobilisieren. Seit der zunehmenden Verschärfung des europäischen Migrations- und Grenzregimes in Folge der EU-Binnenintegration stehen Schutzsuchenden immer weniger legale Einreisemöglichkeiten nach Europa offen. Viele Menschen haben daher auf ihren erzwungenermaßen heimlichen Wegen nach Europa nur die eigene Körperkraft zur Verfügung, um ihre (wenigen) Habseligkeiten über weite Strecken hinweg zu transportieren. Die Art des Gepäckstücks, oder besser: dessen Auswahlmöglichkeit, spiegelt dabei auch die Möglichkeiten bzw. Erfordernisse der Mobilität wider: »Ich hatte nur einen Rucksack mit, da ich wusste, dass ich viel zu Fuß gehen muss auf der Balkanroute« (Österreichisches Museum für Volks-

ers by juxtaposing travel-chest antiquities with a passenger flight bag and with a plastic bag. Chests were popular furnishings across social divides – in aristocratic, middle-class and peasant living spaces. All kinds of things were stored in them, and they often had a secondary function as seats. Chests also accompanied their owners during relocations. They housed and transported a woman's dowry, for example, when she moved into a new household after marriage. Servants' chests held the few belongings these servants owned (Mihm 2001: 18). The travel bag and plastic bag displayed, by contrast, only enclosed what was left from the belongings of the respective refugee who carried that receptacle.

While heavy chests demand an appropriate infrastructure to move them, the modern travel bag and the plastic bag are easier to mobilise. Because of the EU's increasingly stringent migration and border regime on its external perimeter, people seeking refuge have fewer legal options to enter Europe than ever before. Many people must resort to relying on their own physical stamina when transporting their belongings over long distances, when they are forced to follow clandestine routes into Europe. The kind of luggage or the baggage choices they have made reflect both the possibilities and necessities of mobility. As one migrant cited in the exhibition *The Shores of Austria* put it: "I brought only one backpack because I knew that I would have to walk a lot on the Balkan route" (Vienna *Volkskunde-Museum* 2018: 73). Celine, who fled from Nigeria, also reflected in a conversation on what she could take with, her

kunde 2018: 71), ist beispielsweise in der Ausstellung *Die Küsten Österreichs* zu lesen. Auch Celine, die aus Nigeria flüchtete, reflektiert in einem Gespräch die mit unterschiedlichen Fortbewegungsarten verknüpften Möglichkeiten der Mitnahme von Dingen: »Eigentlich hatte ich keine Möglichkeit, etwas mitzunehmen, […] weil ich durch die Wüste gekommen bin, nicht mit dem Flugzeug. Ich meine, mit dem Flugzeug ist es einfacher, man kann Dinge mitnehmen, die sehr wichtig sind. Aber es war mehr wie eine Reise auf der Straße, weißt du?« (Celine L.)

Das jeweilige (Reise-)Gepäck wirkt somit nicht nur begrenzend im Sinne der Dinge, die mitgeführt werden können. Größe und Form des Gepäcks stehen auch in einem engen Zusammenhang mit der Form der Fortbewegung; sie beschränken die Mobilität und ermöglicht sie zugleich. Muss das Gepäck getragen werden oder kann es während der Reise verstaut werden? Wie lange muss es getragen werden, wie nahe ist es dem (mobilen) Körper? Vom Koffer in der Hand über die Schultertasche bis hin zum Rucksack am Rücken: Auch die Nähe zum Körper beeinflusst die Mobilität. Der Rucksack gerät dabei gänzlich aus dem Blickfeld und wird gewissermaßen zur Erweiterung des reisenden Körpers.

Zugleich verrät auch Form und Gestalt des Gepäcks nicht nur etwas über den gesellschaftlichen Status, sondern auch über die damit einhergehenden Mobilitätsmöglichkeiten. »Luggage labels«, also Aufkleber und Gepäckbanderolen, dienen etwa der Sichtbarmachung und Selbstmarkierung als »Weltenbummler«, als kosmopolitische Person, die viele Länder bereist hat und das auch gerne nach außen trägt. Auf der anderen Seite be-

based on the different modes of transport she used: "Actually I didn't have the opportunity to take anything with me … because I came through the desert, not flying. I mean, flying is easier, you can take things that are very important. But it was more like a journey by road, you know?" (Celine L.)

Thus, not only does travel luggage limit what can be carried, the size and shape of the luggage also has a close bearing on the modalities of transport: Luggage restricts mobility and enables it simultaneously. Does an individual have to carry their luggage themselves/keep it on their person, or can they stow it away during the trip? For how long does a case have to be carried? How close will it be, in sheer physical terms, to a mobile body? From a hand-carried suitcase, to a bag carried on one's shoulder, to a rucksack carried on the back, the proximity of baggage to the body also impacts mobility. The backpack disappears out of sight and becomes, so to speak, an extension of the travelling body.

Concurrently, the form and design of the luggage says something not merely about social status but also about the mobility options available to the carrier. Luggage tags are used, for example, to make it easier for their owners to identify their bags, but also to signal that the owner sees themselves as a 'globetrotter', as a cosmopolitan person who has travelled to many countries and wants to advertise this fact. Many refugees, on the other hand, report that they used strategies of invisibility so as *not* to be perceived as fleeing, as this would risk attracting the authorities' attention. These strategies went

richten viele Geflüchtete, dass sie Strategien des Unsichtbarmachens anwandten, um eben nicht als Person auf der Flucht erkannt zu werden und so staatlichen Autoritäten zu entgehen. Zu diesen Strategien zählte nicht nur die Auswahl einer bestimmten Fluchtroute (im Wald, in der Nacht, zu Fuß etc.), sondern ebenso das Wegschmeißen bestimmter Gepäckstücke, wie etwa des Schlafsacks. Strategien des Unsichtbarmachens von Gepäck wird in diesem Zusammenhang zum Unsichtbarmachen (unerwünschter) Mobilität an sich.

Wie Status und (staatlich festgelegte) Mobilitätsmöglichkeiten ineinandergreifen, wird auch in einer von uns besuchten Ausstellung mit dem Titel *Die Wiener in China. Fluchtpunkt Shanghai* deutlich, die 22 jüdische Familien porträtiert, die in Shanghai Zuflucht

beyond the selection of specific routes and means of travel (which included proceeding through forests, at night, and on foot), to encompassing the discarding of certain items of luggage, such as sleeping bags. In this context, strategies of making luggage invisible become central to rendering invisible a whole form of mobility that large sections of the authorities deem undesirable.

How status and hegemonically determined mobility options intertwine is explored in the exhibition *Little Vienna in Shanghai* – which ran from October 2020 to June 2021 and which portrayed 22 Jewish families who found refuge from Nazi persecution in Shanghai. A hat case presented in this exhibition points to the

Abb. 3: Hutkoffer in der Ausstellung *Die Wiener in China. Fluchtpunkt Shanghai* des Jüdischen Museums Wien, 2021. ©Patrizia Gapp

Image 3: Hat case in the exhibition *Little Vienna in Shanghai* in the Jewish Museum Vienna, 2021. ©Patrizia Gapp

vor der Verfolgung durch das Nazi-Regime gefunden haben. So lässt ein in dieser Ausstellung präsentierter Hutkoffer auf die bürgerliche Herkunft seiner Besitzerin schließen und macht damit – für uns durchaus beklemmend – auf einen wesentlichen Aspekt der Flucht aufmerksam: Eine Flucht nach Shanghai war sehr kostspielig und war daher in erster Linie für wohlhabendere Familien der Mittel- und Oberschicht eine Möglichkeit, das nationalsozialistische Österreich hinter sich zu lassen. Neben Shanghai als Ort des Exils lässt sich auf dem Hutkoffer zudem ein früherer Aufenthaltsort seiner Besitzerin ablesen – ein Aufkleber verrät, dass er, aufgegeben mit der Bahn, zuvor als Reisebegleiter für eine

social class of its former owner, thus disconcertingly drawing attention to an essential aspect of their flight. Travelling to Shanghai cost a lot, meaning that escaping Nazi repression in Austria by this route was reserved for wealthy middle- and upper-class families. In addition to Shanghai as place of exile, the hat case also reveals the previous travels of its owner: Given up as baggage on the train, it had previously accompanied its owner on a summer trip to Wörthersee in Austria.

Even if travel and migration are fundamentally different quantities,[5] all people on the move are dependent on the mobility channels available to them. Regarding those Viennese forced to flee to Shanghai, tickets and luggage tags reveal that it was a large, so-called "Aryan" travel company that profited financially from Jews emigrating by rail to China. There is cruel irony in the fact that the same company, the Deutsche Reichsbahn [German Imperial Railways] was also central in the deportation of the Jews and the transportation of forced labourers to the death camps and labour camps – and reaped substantial financial rewards for doing so.[6]

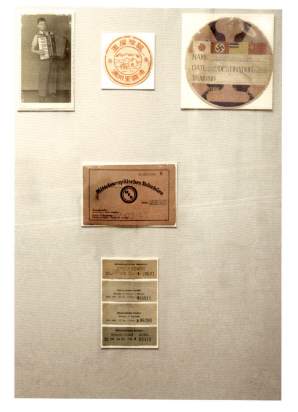

Abb. 4: Fahrkarten in der Ausstellung *Fluchtpunkt Shanghai. Wiener in China* des Jüdischen Museums Wien, 2021. ©Patrizia Gapp

Figure 4: Travel tickets in the exhibition *Little Vienna in Shanghai*, Jewish Museum Vienna, 2021. ©Patrizia Gapp

Abb. 5: Koffer in der Ausstellung *Die Küsten Österreichs* – Volkskundemuseum Wien (Themenraum »Geschichte des Volkes«), 2021. ©Patrizia Gapp

Image 5: Suitcase in the room titled History of the People [Geschichte des Volkes] in *The Shores of Austria* at the Volkskundemuseum Vienna, 2021. ©Patrizia Gapp

Sommerfrische am Wörthersee verwendet worden sein dürfte.

Auch wenn sich Reisen und Migration grundlegend voneinander unterscheiden (siehe Chambers 1994: 5), sind Menschen auf der Flucht auf die Mobilitätskanäle angewiesen, die ihnen zur Verfügung stehen. Im Falle der Flucht nach Shanghai zeigen Fahrkarten und Kofferanhänger, dass es ein »arisches« Reisebüro war, welches finanziell von der Auswanderung der Jüd:innen auf dem Landweg profitierte. Es entbehrt nicht einer grausamen Ironie, dass eben jenes mitteleuropäische Reisebüro auch an den Deportationen von Jüdinnen und Juden und den Transporten von Zwangsarbeiter:innen beteiligt war und dafür Provisionen kassierte.[6]

»Man darf keine Tasche mitnehmen« – Vom Verlust

Immer wieder müssen Menschen auf den komplizierten Routen nach Europa Gepäck zurücklassen. Sei es, weil Dinge sich abnutzen, kaputt oder verloren gehen. Teilweise ändern sich Reiserfordernisse, bestimmte Dinge werden nicht mehr gebraucht und daher entsorgt. Nicht immer geschieht das Zurücklassen von Dingen jedoch freiwillig. Gerade in Bezug auf Mittelmeer-Überquerungen häufen sich die Erzählungen, dass Gepäck oder Teile davon zurückgelassen werden musste: »[...] wir sind mit einem kleinen Boot nach Griechenland durch das Meer ... [Der Schlepper] hat gesagt ist zu viel, 33 Leute in einem kleinen Boot waren wir, er hat gesagt wir dürfen keine Sachen tragen. Wir mussten alles wegschmeißen« (Abdul N.).

Auch die oben erwähnte, in der Ausstellung im Volkskundemuseum präsentierte Reisetasche wurde von der unbekannten Frau, die sie für ihre Flucht nach Europa packte, an einem türkischen Strand zurückgelassen. Die Initiatoren des Projektes fanden diese Tasche in geöffnetem Zustand, und es lag ein noch feuchtes Handtuch darauf – für sie ein Beleg dafür, dass unmittelbar davor eine Überfahrt nach Griechenland stattgefunden hatte. Die Tasche selbst enthielt persönliche Gegenstände. Auch wenn die Tasche samt Inhalt zurückgelassen wurde, knüpfen daran wichtige ethische Fragen an:[7] Wer darf solche an und für sich sehr persönlichen Objekte besitzen? Hat ein Museum das Recht, zurückgelassene und weggeworfene Dinge auszustellen, und falls ja, auf welche Weise soll dies geschehen? Wie steht es um die Dinge, die es aufgrund von musealen oder

"You can't take a bag with you" – On Loss

Countless people have had to leave luggage behind them along the complicated routes to Europe, whether because bags wore out, broke or got lost. Travel conditions change, and sometimes things are no longer needed: Individuals then want to and are sometimes able to get rid of them. Leaving things behind is not always voluntary, however. Stories about having to part permanently from luggage form a prominent part of narratives about crossing the Mediterranean: "We crossed the sea in a small boat to Greece ... [The trafficker] said we were too many, there were 33 of us in a small boat, he said we weren't allowed to bring along any of our things. We had to throw everything away." (Abdul N.)

The travel bag shown in the Volkskundemuseum exhibition was also left behind on a Turkish beach by an unknown woman, who had packed it for her escape to Europe. The exhibition curators found this bag standing open covered by a damp towel, for them evidence that a crossing to Greece had just taken place. The bag itself contained personal items. Even though the bag and its contents were left behind, important ethical questions remain tied up with them.[7] Who has the right to take possession of objects of such a very personal nature? Is a museum entitled to display objects that have been left behind or thrown away, and if so, how should this be done? What are the ethics of dealing with objects that make it to Europe because

künstlerischen Sammelaktionen nach Europa schaffen, aber ihre Besitzer:innen womöglich nicht?

Sich der Ambivalenz des Sammelns solcher Gegenstände durchaus bewusst, haben die beiden Projektverantwortlichen dennoch an der türkischen Küste entschieden, diese Tasche zunächst in den Sammlungsbestand aufzunehmen. Nach ihrer Rückkehr wollten sie die Besitzerin suchen und ihr die Tasche zurückgeben. Sie überlegten, welche Geschichte daraus entwickelt und im Museum gezeigt werden kann. Zurück in Wien scheiterten jedoch alle Bemühungen, anhand der Telefonnummern im Notizheft die Besitzerin der Tasche ausfindig zu machen – wohl auch aufgrund der (berechtigten) Skepsis der Angerufenen, was denn der eigentliche Zweck dieses Anrufs sein könnte. Jetzt ausgestellt, wird die Tasche zu einem Überbleibsel der schwierigen Reise in ein umkämpftes Europa, und sie symbolisiert den Widerstand, die Beharrlichkeit und die Handlungsfähigkeit von Menschen auf der Flucht – trotz der ›offiziell geschlossenen‹ Routen. Aufgegeben als Fluggepäck im Rahmen der Sammlungsaktion erzählt der nicht entfernte Gepäckstreifen auf der Tasche zugleich von unterschiedlichen Transport- und Mobilitätsmöglichkeiten: Während Migrant:innen sich auf oft gefährliche Reisen begeben müssen, können Privilegierte sich frei bewegen und zurückgelassene Gegenstände auf den illegalisierten Routen für europäische Museen einsammeln.[8]

of museum or artistic collection activities, while their owners may not have made it? Well aware of the ambiguities of assembling such artefacts, the curators of *The Shores of Austria* decided nonetheless to incorporate the bag from the Turkish beach into the first stage of the collection. On returning from Turkey to Austria, they intended to look for the owner, return the bag, see what story might emerge from the action and consider how this could be represented in the museum. Their efforts to find the owner, using phone numbers in the notebook discovered in the bag, failed. The people phoned up may have felt justifiably sceptical regarding what purpose the enquiries actually served. Now displayed in the museum, the bag bears witness to a difficult journey to a contested Europe and symbolises the resilience, perseverance and initiative of people forced to go on the move – despite established routes already being classified as 'officially closed'. Handed in as baggage for the flight home from the collection expedition, the luggage tag was left on, as further evidence of different transport and mobility options. While migrants often have to undertake dangerous journeys, the privileged can travel freely and gather relics for European museum collections, using routes, if they wish, that have long since been criminalised for others.[8]

From Suitcase to Flight Baggage

»(Wiener) Kultur im Gepäck«? – Vom Nicht-Ankommen und Dazwischen-Bleiben

Im Kontext bedrängter Lebenssituationen verweisen ausgestellte Koffer zumeist auf den Augenblick, in dem Menschen alles Vertraute ihrer vormaligen Lebenszusammenhänge hinter sich lassen (müssen). Als Fritz Hungerleider mit seiner Familie vor dem Nationalsozialismus nach Shanghai flüchtete, nahm er ein Grammophon und eine schwere Beethoven-Büste mit. Die beiden Objekte stehen in der bereits erwähnten Shanghai-Ausstellung für sein Aufwachsen in einem musikaffinen Wiener Umfeld, in dem er vor der Flucht von einem Philharmoniker Violinunterricht bekam (Pscheiden/Spera 2020: 194). Auch nach der Flucht blieb Fritz seiner musikalischen Leidenschaft treu und trat in Shanghai erstmals als Tenor auf (Pscheiden 2020: 47). Im Mittelpunkt der Ausstellung stehen die vielfältigen Bemühungen der Geflüchteten, sich mit »little vienna«, also »Klein-Wien«, ein eigenständiges Refugium für ihre – quasi im Gepäck mitgebrachte – »Wiener« Kultur und Identität zu schaffen.

Die Präsentation der mitgebrachten Dinge lässt in Ausstellungen mitunter ein idealisiertes Bild entstehen, wonach Migrant:innen sich mit ihrer jeweiligen (National-)Kultur im Koffer aufgemacht hätten, um diese an anderen – ihnen noch fremden Orten – eins zu eins wieder auszupacken. Da erfordert es häufig ein genaues Hinsehen, um jene Erzählfragmente ausfindig zu machen, die sich der Darstellung von einer ungebrochenen kulturellen Kontinuität nach der Migration entziehen. In der Shanghai-Ausstellung finden sich etwa verstreute Hinweise auf die Reaktivierung von Religiosität aufgrund

der Exilerfahrung oder auf Aspekte des Kulturtransfers, wie etwa die Verwendung chinesischer Kerzenständer für den Sabbat oder das Erlernen des chinesischen Mahjong-Spiels. Fritz Hungerleider wandte sich in Shanghai dem Buddhismus zu. Er ging 1947 nach Österreich zurück und gründete zwei Jahre später die Buddhistische Gesellschaft in Österreich, deren Präsident er für mehr als zwei Jahrzehnte war.[9]

Die Vorstellung, dass der Koffer ein Container für Dinge (und damit Ideen) ist, die an andere Orte transportiert werden, ausgepackt, verwendet, wieder (unverändert) eingepackt und »nach Hause« oder an andere Orte gebracht werden, ist stark vereinfachend. In Erzählungen zum aktuellen Fluchtgeschehen wird deutlich, dass die Flucht kein klar abgrenzbares und lineares Ereignis darstellt, bei dem Menschen einen Ort verlassen und an einem anderen ankommen: Reza beginnt die Erzählung der eigenen Flucht aus dem Iran mit der Fluchtgeschichte seines Vaters, der als junger Mann aus Afghanistan in den Iran fliehen musste – das Land, in dem Reza später geboren wurde und aufgewachsen ist. Als Angehörige einer unterdrückten Minderheit – der Hazara – sahen sie sich dort staatlichen Repressionen ausgesetzt, die es – diesmal für die gesamte Familie inklusive der nachfolgenden Generation – notwendig machte, das Land (wieder) zu verlassen. Selbst die eigene Flucht stellte im Fall von Reza ein mehrjähriges Ereignis mit vielen Zwischenaufenthalten und ungewissem Ausgang dar: »Wir haben nichts gewusst. Wir haben weder gewusst, wie lange wir warten müssen, noch haben wir gewusst, ob das wirklich funktioniert ... Wir hatten viele Hoffnungen, dass es funktionieren wird, aber gewusst haben wir das nie« (Reza S.).

has taken place. In the Shanghai exhibition, scattered references also exist to a reactivation of religious practice during the exile experience and to aspects of cultural transfer, such as using Chinese candelabras for the Sabbath or learning to play the Chinese game Mah-jong. Fritz Hungerleider, for example, became involved in Buddhism in Shanghai. Two years after returning to Austria in 1947, he founded the Buddhist Society in Austria, of which he was president for more than two decades.[9]

The concept that a suitcase is a container for things – and thus ideas – that are transported to other places, unpacked, used and repacked but remain unchanged, before finally being brought 'home' or to some place, will not suffice. Contemporary narratives about flight, for example, make evident that this is anything but a clearly defined and linear event, in which an individual leaves one place and arrives at another. Reza begins the story of his own flight from Iran with the migration story of his father, who as a young man had to flee from Afghanistan to Iran, the country where Reza was later born and grew up. However, as members of an oppressed ethnic minority – the Hazaras – his family also fell victim to state repression in Iran, making it necessary to flee from a second state – this time as a whole family. Reza experiences his migration as an event lasting several years, with many interim stops and an uncertain outcome: "We didn't know anything. We didn't know how long we would have to wait, we didn't know whether it would really work ... We had high hopes that it would work, but we were never sure." (Reza S.)

Das in diesem Zusammenhang erzwungene Leben im Provisorium und das Nichtankommen im Kontext von Flucht und Vertreibung zeigt sich auch im Volkskundemuseum. Hier lässt sich anhand einer ausgestellten Plastiktüte ermessen, was es für Menschen bedeutet, dass das gegenwärtige Asylregime sie nicht ankommen lässt – wenn sich also die Flucht auch ohne die eigentliche physische Bewegung im alltäglichen Leben fortsetzt. Der Begleittext verrät folgendes über dieses Objekt:

> Wozu braucht man einen Schrank, wenn man damit rechnen muss weiter zu müssen – oder abgeschoben zu werden? Ein Flüchtender berichtet, den Schrank in seiner Unterkunft nicht genutzt zu haben. Ohne gültigen Aufenthaltstitel sei die Flucht nicht vorbei und das Fluchtgepäck stellte für ihn daher die sinnvollste Verwahrungsmöglichkeit dar. (Österreichisches Museum für Volkskunde 2018: 71)

Der Schrank wird auf diese Weise zum Sinnbild für das Bleiben und sich Niederlassen(können) – ein fester verankertes Haus für aufzubewahrende Dinge, das Menschen ohne Status vorenthalten wird.

The Volkskundemuseum exhibition demonstrates that such political and social constellations force people into living in provisional, material conditions and also explores the accompanying psychological state of not having arrived. The exhibited plastic bag communicates what it means to people that the current migration regime does not permit them to arrive – when the feeling of fleeing continues in everyday life, even after actual physical movement has ended. The accompanying text reveals the following about this object:

> What is the use of a wardrobe, when one expects to move along or be deported anyway? A refugee reports that he did not use the locker in his accommodation. Without a valid residence permit he is still on the run, therefore the luggage used for escape is still the most useful storage option, he tells us. (Vienna Volkskundemuseum 2018: 219).

The wardrobe thus becomes a symbol for remaining and settling down – a way of storing things in the kind of fixed abode that is normally denied to people without legal residency status.

Schlussbemerkung: Konstante Veränderung und verändernde Konstanz

Der Beitrag könnte an dieser Stelle mit dem Ankommen und den Praktiken des Auspackens, des Sicheinrichtens und des Beheimatens enden, womit die suggerierte Linearität des Koffers auf eine Linearität in der Erzählung trifft und die Ordnung wiederhergestellt wäre. Mit dem Nichtankommen, dem Dazwischenbleiben schon benannt, bedeutet jedoch Flucht auch eine Zäsur und damit Veränderung. So beschreibt Reza über die Mitnahme oder das Zurücklassen von Dingen, dass er die Flucht nicht nur als langwierigen Prozess wahrgenommen hat. Er erzählt, metaphorisch verdichtet, dass die Fluchterfahrung gleichsam eine Veränderung darstellt, die nicht nur äußerlich ist, sondern sich auch auf das eigene Selbstverständnis auswirkt:

> ... Du entscheidest nicht nur, welche physikalischen Teile du mitnimmst oder zurücklässt ... Du nimmst oder lässt auch Teile deines Menschen, halt deines Charakters zurück. Es ändert sich so viel auf der Flucht ... Du nimmst so vieles mit, das du nie hattest. Du lässt so vieles zurück, das du vielleicht nie wieder bekommst, als Teil deiner Seele, Teil deines Charakters. (Reza S.)

Auch das in den Migrationsausstellungen präsentierte Gepäck hält Menschen nicht immer in Zwischenräumen oder markiert das »Ende einer Reise«. Im Volkskundemuseum lässt sich das Zeigen des Koffers eines in den 1990er Jahren angekommenen Flüchtlings als

Closing remarks: Constant Change and Changing Constants

This essay could end here: with arriving and the practice of unpacking, settling in and making yourself at home, the straight lines of the modern suitcase mirroring the linearity in the narrative. Order would be restored. Yet, as already proposed through the status and concepts of not-yet-having-arrived and remaining in-between, flight also means breaking with the past and change. Thus, when Reza describes taking things with him or leaving them behind, his perception goes beyond acknowledging that flight is a protracted process. Using compact metaphors, he narrates how the experience of fleeing can be equated with a change that is more than merely external, and indeed even changes his self-understanding:

> ... You don't just decide which physical things you take with you or leave behind ... You also take or leave behind parts of yourself as a human, like of your character. So much changes while fleeing ... You take so much with you that you never even had. And you leave so much behind that you'll maybe never get back again, as a part of your soul and a part of your character. (Reza S.)

Even the luggage presented in the two migration exhibitions discussed does not mark people as being in intermediate spaces or at the 'end of a journey'. In the Volkskundemuseum, the suitcase of a

Versuch der Kurator:innen interpretieren, die biographischen Erfahrungen von Geflüchteten in die »Geschichte des Volkes« einzuschreiben. Eingebettet zwischen zwei volkskundlichen Objekten mit offensichtlichem Habsburg-Bezug, thematisieren die Projektbeteiligten mit dem Koffer das Spannungsverhältnis zwischen der migrantischen Anwesenheit und den volkskundlichen Sammlungen, die den Blick vornehmlich auf das – wie auch immer imaginierte – kollektive »Eigene« richten.

Eine beliebte Spielart des Koffers in Migrationsausstellungen ist die Zurschaustellung seines Inhaltes.[10] In gängigen Inszenierungen sind es dann zumeist besondere bzw. exotisch bunte Dinge aus dem Gepäck der Zugewanderten, welche die Vielfalt der Kulturen verdeutlichen sollen. Es wird auf diese Weise von einer klaren Unterscheidbarkeit von einheitlichen ethno-nationalen Kulturen ausgegangen, die nebeneinander existieren. Solche (Re-)Präsentationsweisen befördern »die Vorstellung von Migranten als Zusatz, gar Dekoration einer im Kern unverändert und homogen gedachten Gesellschaft« (Baur 2009: 20).

Indem der Koffer im Volkskundemuseum geschlossen bleibt und ihm so kein mitgebrachtes kulturelles Gepäck entspringt, werden auch sich konstant in Veränderung befindliche Selbstverständnisse und Zugehörigkeiten sichtbar. Wenn an anderer Stelle in diesem Raum zu lesen ist »Es gibt viele Erzähler. Ich möchte einer von ihnen sein« (Österreichisches Museum für Volkskunde 2018: 107), dann löst der Koffer gemeinsam mit den anderen Objekten der Schausammlung diesen berechtigten Anspruch der Neuhinzukommenden nach Anerkennung ihrer eigenen

refugee who arrived in the 1990s can be interpreted as an attempt by the curators to inscribe biographical experiences of refugees into the "history of the Austrian people". Sandwiched between two anthropological objects with a rather obvious link to the Habsburg Empire, the curators use this case to address the tension between migrant presences and ethnological collections that focus primarily on a collective sense of what is 'our own' – however that is specifically imagined.

A popular play on the suitcase in migration exhibitions is the straightforward display of its contents. In common stagings, it is typically special or exotically colourful things from the immigrants' luggage that are supposed to illustrate the diversity of the world's cultures. The starting point of such an approach are uniform, ethno-national cultures that can be clearly distinguished from one another and that exist side by side.

Such modes of (re)presentation foster "the notion of migrants as an addition, or even as a decoration for a society that remain unchanged, at its core, and that is conceived of in homogeneous terms" (Baur 2009: 20).

Keeping the suitcase closed and preventing any cultural baggage from escaping[10] in fact reveals constantly changing self-perceptions and associations. Because the following can be read elsewhere in the same room – "There are many storytellers. I would like to be one of them" (Vienna Volkskundemuseum 2018: 107) – the suitcase fulfils the justified claim of the newcomers for recognition and for inter-

Deutungshoheit über ihre (Flucht-)Erzählungen ein. An dieser Stelle sei noch einmal Rezas Beschreibung aufgegriffen, die im Kleinen zeigt, was auch für das Große gilt: Sie verweist über die unmittelbare persönliche Erfahrung hinaus auf das (historisch immer schon gegebene) Spannungsfeld von Ankommen, Mobilität und Veränderung in einer (damit immer schon) postmigrantischen Welt.

pretative sovereignty over their own narratives – whether these are about fleeing or about other subjects. This takes us back to Reza's description, which elucidates on a small scale what is valid on a larger one. His words point beyond direct personal experiences to the area of conflict between arriving, mobility and change. These tensions have always existed in history in what, thus, has always been a postmigration world.

Literatur | References

Baur, Joachim. 2009. *Die Musealisierung der Migration, Einwanderungsmuseen und die Inszenierung der multikulturellen Nation*. Bielefeld.

Baur, Joachim. 2009. Flüchtige Spuren – bewegte Geschichten. Zur Darstellung von Migration in Museen und Ausstellungen. In: *Tagungsdokumentation: Inventur Migration*. Oberhausen.

Chambers, Ian. 1994. *Migrancy, Identity, Culture*. London.

Hamilakis, Yannis. 2019. Planet of Camps: Border Assemblages and Their Challenges. *Antiquity* 93 371:1371–1377.

Harlan, Susan. 2018. *Luggage*. New York.

Mertus, Julie; Tesanovic, Jasmina; Metikos, Habiba; Boric, Rada. 1997. *The Suitcase: Refugee Voices from Bosnia and Croatia*. Berkeley

Mihm, Andrea. 2001. *Packend … Eine Kulturgeschichte des Reisekoffers*. Marburg.

Pscheiden, Daniela. 2020. Little Vienna in Shanghai. In: Pscheiden/Spera, *Die Wiener in China: Fluchtpunkt Shanghai*. Wien, 240–251.

Österreichisches Museum für Volkskunde. 2018. *Die Küsten Österreichs: Ausstellungskatalog*. Wien.

Selheim, Claudia. 2019. Koffer – keineswegs nur für die Reise, In: Claudia Selheim/Ulrich G. Großmann, *Reisebegleiter – mehr als nur Gepäck*. Heidelberg.

Soto, Gabriella. 2018. Object Afterlives and the Burden of History: Between »Trash« and »Heritage« in the Steps of Migrants. *American Anthropologist* 120 (3):460–473.

Wonisch, Regina. 2020. Zur Repräsentation von Migration in Ausstellungen: Aktuelle Entwicklungen in Österreich, *ÖZG* 1:224–238.

Vienna Volkskundemuseum. 2018. *The Shores of Austria: The Latest Display Collection of the Austrian Museum of Folk Life and Folk Art*. Wien.

Anmerkungen

1 Das von uns zusammen mit Johannes Pflegerl, Dieter Bacher, Richard Wallenstorfer und Barbara Stefan bearbeitete Thema »(Nicht) im Gepäck?« ist einer von sechs Teilbereichen des Projektes »Mobile Dinge, Menschen und Ideen. Eine bewegte Geschichte Niederösterreichs« (https://www.mobiledinge.at/), das unter anderem am Ilse Arlt Institut für Soziale Inklusionsforschung bearbeitet wird.

2 Daniela Pscheiden (Kuratorin): *Die Wiener in China. Fluchtpunkt Shanghai*, Jüdisches Museum Wien, 21.10.2020–27.6.2021 (http://www.jmw.at/de/exhibitions/die-wiener-china-fluchtpunkt-shanghai).

3 Die Dauerausstellung wurde im Rahmen des Projekts »Museum auf der Flucht« überarbeitet, siehe »Die Küsten Österreichs. Die neue Schausammlung des Volkskundemuseum Wien«, https://www.volkskundemuseum.at/diekuestenoesterreichs

4 Angeführte Namen im Artikel sind anonymisiert.

5 Das Wort »besitzen« kommt ursprünglich von *auf etwas sitzen*, dies erwähnte Dagmar Czak (Vermittlerin im Volkskundemuseum) im Rahmen einer informellen Führung im Volkskundemuseum am 30.10.2020. Vgl. auch: »besitzen«, in: Wolfgang Pfeifer et al., *Etymologisches Wörterbuch des Deutschen* (1993), digitalisierte und von Wolfgang Pfeifer überarbeitete Version im *Digitalen Wörterbuch der deutschen Sprache*, https://www.dwds.de/wb/etymwb/besitzen, abgerufen am 14.9.2021.

6 Sven Felix Kellerhoff: »Zwei Pfennig pro Kopf und Bahnkilometer ins KZ«, *Die Welt* vom 3. März 2013 (https://www.welt.de/geschichte/zweiter-weltkrieg/article113200916/Zwei-Pfennig-pro-Kopf-und-Bahnkilometer-ins-KZ.html).

7 Vgl. dazu auch die Fragen diskutiert von Yannis Hamilakis: Planet of Camps: Border Assemblages and Their Challenges, in: Antiquity 93 (371), 2019, 1371–1377; hier S. 1376; Gabriella Soto: Object Afterlives and the Burden of History: Between »Trash« and »Heritage« in the Steps of Migrants, in: *American Anthropologist* 120 (3), 2018, S. 460–473.

8 Diese Aspekte wurden am 13.9.2020 von den Au-

Notes

1 The theme we worked on together with Johannes Pflegerl, Dieter Bacher, Richard Wallenstorfer and Barbara Stefan "(Nicht) im Gepäck?" is one of six parts of the project "Mobile Dinge, Menschen und Ideen. Eine bewegte Geschichte Niederösterreichs", https://www.mobiledinge.at/ a collaboration between the Ilse Arlt Institute for Social Inclusion Research, Austria and other partners.

2 Daniela Pscheiden, curator, *Little Vienna in Shanghai*, Jewish Museum Vienna, 21.10.2020–27.6.2021. (http://www.jmw.at/de/exhibitions/die-wiener-china-fluchtpunkt-shanghai).

3 The permanent exhibition was renewed as part of the project "Museum on the Move" [Museum auf der Flucht]. See *The Shores of Austria. The Latest Display Collection of the Austrian Museum of Folk Life and Folk Art*. Showing since 19.9.2018. https://www.volkskundemuseum.at/theshoresofaustria

4 All names in this paper are pseudonyms.

5 On the difference see, for example, Chambers, *Migrancy, Identity, Culture*. London 1994 p. 5

6 Sven Felix Kellerhoff, "Zwei Pfennig pro Kopf und Bahnkilometer ins KZ" [Two Pfennigs per Head and per Rail Kilometre], *Die Welt*, 3 March 2013.

7 See the questions asked on this issue by Yannis Hamilakis, "Planet of Camps: Border Assemblages and Their Challenges", *Antiquity* 93 371, 2019, 1371–1377, p. 1376; Gabriella Soto, "Object Afterlives and the Burden of History: Between 'Trash' and 'Heritage' in the Steps of Migrants", *American Anthropologist* 120(3), 2018, pp. 460–473.

8 These aspects were discussed during a tour of the exhibition with one of the curators, Alexander Martos, on 13.9.2020; cf. the critical discussion in: Regina Wonisch, "Zur Repräsentation von Migration in Ausstellungen: aktuelle Entwicklungen in Österreich", *ÖZG* 1, 2020, 224–38.

9 Stephan Löwenstein: Eine Frankfurterin in Wien: "Gerade noch habe ich das Schiff erreicht", *FAZ.net*, 17 August 2014, https://www.faz.net/aktuell/gesellschaft/menschen/eine-frankfurterin-in-wien-gerade-noch-habe-ich-das-schiff-erreicht-13098266.html

10 For more detail on the practice of opening suitcases in exhibitions, see: Joachim Baur,

torinnen im Rahmen einer Ausstellungsführung mit Alexander Martos als einem der Kuratoren diskutiert, vgl. kritisch reflektierend Regina Wonisch: Zur Repräsentation von Migration in Ausstellungen: Aktuelle Entwicklungen in Österreich, ÖZG 1, 2020, S. 224–238

9 Stephan Löwenstein: Eine Frankfurterin in Wien: »Gerade noch habe ich das Schiff erreicht«, *FAZ.net*, 17. August 2014, https://www.faz.net/aktuell/gesellschaft/menschen/eine-frankfurterin-in-wien-gerade-noch-habe-ich-das-schiff-erreicht-13098266.html

10 Vgl. ausführlicher zur Öffnung von Koffern in Ausstellungen Joachim Baur: Flüchtige Spuren – bewegte Geschichten. Zur Darstellung von Migration in Museen und Ausstellungen, in: *Tagungsdokumentation: Inventur Migration*, Oberhausen 2009, S. 14–26, hier 20f.

Flüchtige Spuren – bewegte Geschichten. Zur Darstellung von Migration in Museen und Ausstellungen. In: *Tagungsdokumentation: Inventur Migration*, Oberhausen 2009, p. 14–26, here 20–21.

Die Papierspur der Migration
Fragmente einer Ethnographie des Reisepasses

The Paper Trail of Migration
Fragments of an Ethnography of the Passport

Romm Lewkowicz

Aksaray, Istanbul, 2017

Bilal[1] ist aufgeregt. Zum ersten Mal seit langem hat er einen Abend frei und wird einige Freunde zum Abendessen zu Besuch haben. Er macht Mandi, ein jemenitisches Gericht aus Reis und Hühnchen. Mandi ist eines der vielen Gerichte, die Bilal regelmäßig in einem jemenitischen Restaurant im Istanbuler Migrantenviertel Aksaray zubereitet, wo er als Koch arbeitet. Da er sich eine Wohnung mit drei anderen Geflüchteten teilt, hat er kaum Platz für Gäste. Doch in dieser Woche sind zwei seiner Mitbewohner – Bola aus Nigeria und Yusuf aus Eritrea – nach Izmir aufgebrochen, wo sie versuchen werden, das Meer in Richtung einer der Ägäis-Inseln zu überqueren. Bilal freut sich, dass er heute ein bisschen mehr Platz hat. Allerdings macht er sich auch Sorgen, wie er die Miete allein bezahlen soll. Es ist inzwischen zu einer Routine geworden: Die Mitbewohner:innen schwören anfangs, dass sie bleiben wollen, um dann ohne jede Vorankündigung nach Izmir aufzubrechen. Bilal hatte auch selbst einmal versucht, über die Meeresroute aus der Türkei rauszukommen. Sein Schlauchboot kenterte fast, schon halb am Ziel,

Aksaray, Istanbul, 2017

Bilal[1] is looking forward to the evening to come. He has got a night off work for the first time in a long while, and he's having some friends over for dinner. He is making Mandi, a rice and chicken dish from his home country of Yemen. Mandi is among the dishes Bilal prepares regularly at a Yemeni restaurant down the street in Aksaray, an Istanbul district with a high migrant population, where he works as a cook. Because Bilal shares an apartment with three other refugees, there is barely any room for guests. But earlier this same week, two of his roommates – Bola from Nigeria and Yusuf from Eritrea – left for the coastal city of İzmir, from where they are aiming to cross the sea towards the Aegean Islands. Bilal is happy about having a little extra space tonight but also anxious about how he is going to pay the rent by himself. This has become a routine: Roommates first swear that they will stick around and then depart for Izmir without giving Bilal any notice. Bilal tried the sea route of getting out of Turkey himself

aber noch in türkischen Gewässern. Er wurde verhaftet und konnte sich nur der Abschiebung entziehen, indem er vorgab, ein syrischer Geflüchteter zu sein (Syrer:innen hatten zu dieser Zeit einen besonderen Schutzstatus in der Türkei). Viele seiner Bootsgefährt:innen blieben in Izmir und versuchten weiter ihr Glück. Von dieser Erfahrung gezeichnet und nun mittellos, beschloss Bilal, seinen europäischen Traum aufzugeben. Zumindest erst einmal.

Während Bilal in seiner Behelfsküche zwischen den Töpfen herumjongliert, erhält er einen Videoanruf von Ali, einem guten Freund, der von der griechischen Insel Chios aus anruft, einem ›Hotspot‹ für Geflüchtete. Sie hatten sich in dem Istanbuler Restaurant getroffen, in dem Bilal noch immer arbeitet. Als sie feststellten, dass sie aus derselben Provinz stammen, schlossen sie trotz Altersunterschieds (Ali ist 15 Jahre jünger) sofort Freundschaft. Bilal richtet die Kamera auf das Mandi, um es Ali zu zeigen, während Ali antwortet, indem er Bilal die Sonne zeigt, die über dem Hafen von Chios auf der gleichnamigen griechischen Insel untergeht. Ihr Videotelefonat läuft während der gesamten Zeit, in der Bilal gemeinsam mit seinen Gästen zu Abend isst.

Später am Abend, nachdem der letzte Gast gegangen ist, eilt Bilal zu seinem Bett, auf dem die Gäste an diesem Abend gesessen haben. Er zieht einen kleinen Koffer hervor, der mit einem Schloss verschlossen und mit einer Tischdecke abgedeckt ist. Er zeigt mir seinen Pass, der neben Alis im Koffer liegt. Das seien die beiden wertvollsten Besitztümer in seiner ganzen Wohnung, sagt er. Es beruhigt ihn, dass beide noch da sind. Ali ließ seinen Pass bei Bilal, bevor er sich auf den Weg über die Ägäis machte. Er befürchtete, dass wenn türkische,

once. His dinghy almost capsized, halfway through the crossing, but still in Turkish waters. He was detained and only managed to evade deportation by pretending to be a Syrian refugee: Syrians had temporarily been granted better protection rights in Turkey at the time. Many of his companions from the dinghy remained in İzmir, to continue to try their luck. Left bruised and penniless by the experience, Bilal decided to shelve his European dream. For the time being, at least.

While Bilal is juggling pots around in his flat's makeshift kitchen, he receives a video call from a good friend, Ali, who is phoning from the Greek island of Chios: a refugee 'hotspot'. They got to know each other at the same Istanbul restaurant where Bilal is still employed. Realizing they were from the same province, they immediately bonded, despite their age gap – Ali is 15 years younger. Bilal points the camera at the Mandi to show it to Ali, while Ali responds by showing Bilal the sun starting to set over the port of Chios on the Greek island of the same name. Their video chat runs on through the entire time Bilal and his guests take to dinner.

Later that night, after the last guest has left, Bilal rushes to his bed on which the guests had sat that evening. He pulls out a small suitcase, which is sealed with a lock and covered with a tablecloth. He shows me his passport, which is in the suitcase alongside Ali's. These are the two most precious possessions in his whole house, he says. It relieves him to see they are both still there. Ali left his passport with Bilal before leaving to cross the

griechische oder Frontex-Küstenwachen ihn auf seiner Reise erwischen würden, diese seinen Pass verwenden könnten, um ihn abzuschieben. Wenn es Ali eines Tages gelingt, den Flüchtlings-Hotspot zu verlassen, wird er Bilal bitten, ihm den Pass nachzuschicken. Es ist wahrscheinlich, dass er ihn in einem künftigen Asylverfahren als Beweismittel brauchen wird.

Bilal sagt, dass er Ali gerne mit dem Pass hilft. Das Bewachen eines Passes sei eine enorme Verantwortung, gibt er zu, eine Tatsache, die ihm nicht so ganz bewusst war, als er Ali diese Idee vorschlug. Solange er im Besitz von Alis Pass ist, kann Bilal Istanbul nicht verlassen, was auch bedeutet, dass er derzeit nicht versuchen kann, per Boot zu fliehen. Wenn die Polizei ihn verhaftet und zwei Pässe entdeckt, könnte sie ihn verdächtigen, ein Schmuggler zu sein. Deshalb ist er ständig in Sorge, Alis Pass zu verlieren, obwohl er sich eingesteht, dass er sich nicht so viele Sorgen machen sollte. »Warum sollte jemand einen jemenitischen Pass stehlen wollen? Er ist völlig nutzlos!«, lacht er. Aber er sorgt sich um Ali, und das bedeutet, dass er sich um das einzige greifbare Überbleibsel kümmern muss, den er in Alis Abwesenheit von ihm hat: seinen Reisepass. Die Bewahrung von Alis Pass gibt Bilal das Gefühl, dass er sich selbst und den Behörden versichern kann, dass er »Papiere hat«. Und auch Ali hat weiterhin das Gefühl, dass er »Papiere« hat, auch wenn sie derzeit physisch von ihm getrennt sind.

———

Im Englischen steht der Begriff ›paper trail‹ (Spur der Dokumente) für die Vorstellung, dass die Meinungen oder Handlungen einer Person durch die Verfolgung einer Kette von Dokumenten nachvollzogen oder aufgedeckt werden können. Die Idee ist, dass Papiere wie

Aegean Sea. He was afraid that if Turkish, Greek or Frontex coastguards caught him during his journey, they would use his passport to deport him. He was also worried that his passport would be damaged or lost during the perilous sea crossing. One day, if Ali manages to leave the refugee hotspot, he will ask Bilal to send it back to him. It's probable that he will need it as evidence during any future asylum case.

Bilal says he is happy to help Ali with the passport. Guarding a passport is an enormous responsibility, he admits, a fact he hadn't fully grasped when he suggested the idea to Ali. As long as he has possession of Ali's passport, Bilal cannot leave Istanbul, which also means he can't attempt to flee by boat at present. If the police arrest him and discover two passports, they may suspect that he is a smuggler. This leaves him continually anxious about losing Ali's passport, even though he acknowledges that he shouldn't be. "Why would anyone want to steal a Yemeni passport? It's utterly useless!" he laughs. But he cares about Ali, and that means having to care for the only tangible trace he has of him in his absence: his passport. Safeguarding Ali's passport makes Bilal feel that he can say to himself and the authorities that he "has papers". And Ali also continues to feel that he "has papers", even if they are currently physically detached from him.

———

In English, the term 'paper trail' represents the idea that a person's opinions or actions may be traced or uncovered by tracking

Fingerabdrücke sind, ein Stück von uns selbst, das wir entgegen unseren bewussten Absichten wie einen klebrigen Schatten hinterlassen. Wie bei Fingerabdrücken folgen Ermittler:innen der Spur von Dokumenten, um eine Person mit einem bestimmten Ort und einer bestimmten Zeit in Verbindung zu bringen, in der Regel als Mittel, um sie zu belasten.

Aber für die Geflüchteten, die die Erfahrung von erzwungener Migration machen, sind Papiere weder ein Schatten noch eine Nebensache. Ihre Anwesenheit wird instinktiv gespürt. Sie sind wertvoll, prekär und offenkundig unbeständig. Weit davon entfernt, eine anonyme oder neutrale Darstellung eines Individuums zu sein, stehen sie für ein Bündel von Beziehungen und Verwandtschaften, die Personen, Verwandte, Reisegefährten, Staaten, Schmuggler, Heimatländer, globale Märkte, Unvergessliches aus der Vergangenheit und Wünsche für die Zukunft miteinander verbinden und verknüpfen.

Die Wege von Migrant:innen und ihren Papieren verzweigen sich immer wieder. Auf ihrer Reise nach Europa müssen sich Migrant:innen und ihre Papiere oft trennen – manchmal, um sich zu einem späteren Zeitpunkt wieder zu treffen. Wie die Geschichte von Bilal zeigt, übernehmen Migrant:innen gelegentlich die Verantwortung für die Reise der Papiere einer anderen Person. Migrant:innen wie Ali und Bilal müssen immer wieder Entscheidungen über ihre eigene physische Reise und die ihrer Dokumente treffen. Ein und dasselbe Papier kann dessen Besitzer:in in einem Moment vor der Abschiebung bewahren oder die Registrierung in einem Lager ermöglichen – während es im nächsten Moment ein erhebliches Risiko darstellt. Die Unzuverlässigkeit der Papiere, die sowohl den Schutz als auch die Abschie-

a chain of documents. The idea is that papers are like fingerprints, a piece of ourselves that we leave behind us unwillingly, contrary to our conscious intentions, like a sticky shadow. As with fingerprints, investigators follow paper-trails to pin down a person to a certain place and time, usually as a means to incriminate them.

But in refugees' experiences of forced migration, papers are neither shadows nor afterthoughts. Their presence is felt viscerally. They are precious, precarious and notoriously volatile. Far from being an anodyne or neutral representation of an individual, they stand for a cluster of relationships and kinships, connecting and tying selves, relatives, travel companions, states, smugglers, homelands, global markets, haunting pasts and desired futures to each other.

The trails of migrants and their papers fork recurrently. On their journey to Europe, migrants and their papers often have to part company – sometimes to reconnect at a later juncture. As Bilal's story demonstrates, migrants occasionally become responsible for the journey of another person's papers. Migrants like Ali and Bilal must reach decisions about their own physical journeys and their paper-trails at every turn. The very same paper can, at one moment, save its holder from deportation or enable registration at a camp – while at the next it poses a substantial risk. The fickle quality of papers to facilitate both protection *and* deportation is characteristic of how the EU's asylum apparatus is intertwined with, and embedded in, the EU's deportation re-

bung bedeuten können, ist bezeichnend für die Verflechtung des EU-Asylsystems mit dem darin eingebetteten EU-Abschiebesystem. Da ihre Wege von denen ihrer Dokumente abweichen, stehen Migrant:innen vor einem wiederkehrenden Dilemma, das kurzfristige Strategien erfordert: Papiere können sich ständig verändern, können von einem Risiko zu einem Vorteil werden und umgekehrt. Migrant:innen stehen vor der fast unlösbaren Herausforderung zu entscheiden, wie, wann und wo sie sich von ihren Papieren trennen, sie bewachen oder wieder mit ihnen in Kontakt treten sollen. Oder mit den Papieren von anderen.

Seit der Entstehung des Nationalstaates steht die Verwaltung von Papieren im Mittelpunkt der verschiedenen Erfahrungen mit erzwungener Migration. In seinem Werk *Flüchtlingsgespräche*, das er nach seiner Flucht aus Nazideutschland im Exil verfasste, schrieb Bertolt Brecht »der Paß ist der edelste Teil von einem Menschen« (Brecht [1940] 2019: 8). Vor dem Hintergrund eines beispiellosen massenhaften Entzugs der Staatsbürgerschaft und der Entstehung von Staatenlosigkeit wurde der Pass zur greifbaren Manifestation dessen, was Hannah Arendt als »das Recht, Rechte zu haben« bezeichnete (1955: 444); ohne ihn sind die Menschen lediglich auf das »nackte Leben« (Agamben 1998) reduziert und der willkürlichen Herrschaft der Polizei (Arendt 1955: 433) ausgesetzt. Sowohl Arendt als auch Brecht mussten beim Grenzübertritt sorgfältig verhandeln, um Papiere zu erhalten und mit Bürokraten in Kontakt zu treten. Brecht und seine Frau Helene Weigel durchliefen die Nervenprobe der parallelen Visumsbeantragung in den USA und Mexiko, während sie vor den Nazis über Dänemark, Schweden und

gime. As their paths continue to diverge, migrants face temporal dilemmas that require temporal strategies: Papers can change constantly from being a risk to being an asset, and migrants face the nearly impossible dilemma of deciding how, when and where to separate, guard or reconnect with their papers. Or with the papers of others.

Since the rise of the nation-state, paper management has been at the heart of diverse experiences of forced migration. In his *Refugee Conversations*, written in exile after fleeing Nazi Germany, Bertolt Brecht wrote that "the passport is the noblest part of a human being (Brecht [1940] 2019: 8)". Against a backdrop of an unprecedented mass stripping of citizenship and the emergence of statelessness, the passport has become the tangible manifestation of what Hannah Arendt termed "the right to have rights" (1951: 296); Without it, people are reduced merely to "bare life" (Agamben 1998) and are exposed to the "arbitrary rule of police decree" (Arendt 1951: 290). Both Arendt and Brecht had to negotiate carefully to acquire papers and liaise with bureaucrats while crossing borders. Brecht and his wife, Helene Weigel, went through the ordeal of parallel US and Mexico visa application procedures, while fleeing the Nazis through Denmark, Sweden and Finland, where they finally got hold of US visas. Arendt was stripped of her German citizenship, failed to acquire even the Nansen passport for stateless persons and was ultimately detained at Gurs internment camp in France. Yet she managed to escape from Gurs, her

schließlich Finnland flohen, wo sie endlich US-Visa erhielten. Arendt wurde die deutsche Staatsbürgerschaft aberkannt, sie erhielt nicht einmal den Nansen-Pass für Staatenlose und wurde schließlich im Internierungslager Gurs in Frankreich inhaftiert. Dennoch gelang es ihr, aus Gurs zu fliehen, wobei ihre Flucht sowohl durch die von ihr vorgelegten Papiere als auch durch die Grenzen, durch die sie sich ›stehlen‹ konnte, bestimmt wurde: von Montauban über Marseille, Portbou und Lissabon zur Einwanderungsbehörde auf Ellis Island in New York. Dort wurde ihr schließlich das Papier der Passlosen ausgehändigt: eine Eidestattliche Erklärung anstelle eines Ausweises. Sie lautete: »Ich möchte dieses Dokument anstelle eines Reisepasses verwenden, den ich als staatenlose Person zur Zeit nicht erhalten kann.«[2]

> Der Paß ist der edelste Teil von einem Menschen. Er kommt auch nicht auf so einfache Weise zustand wie ein Mensch. Ein Mensch kann überall zustandkommen, auf die leichtsinnigste Art und ohne gescheiten Grund, aber ein Paß niemals. Dafür wird er auch anerkannt, wenn er gut ist, während ein Mensch noch so gut sein kann und doch nicht anerkannt wird.
>
> Bertolt Brecht, *Flüchtlingsgespräche* [1940] 2019: 8

Brechts Bild des Passes als ›Körperteil‹ wurde buchstäblich als Vision für das Migrant:innenmanagement in der Schengen-Ära übernommen. Der Pass, der einst als Markenzeichen der Moderne galt, wird in dieser Sichtweise zu einem veralteten Relikt, das durch biometrische Daten ersetzt werden soll.

manner of flight structured by both the paper-trail she had created, and the borders she managed to 'steal' through: from Montauban to Marseille, Portbou, Lisbon and then to the Ellis Island immigrant inspection station in New York. There, at last, she was issued with the paper of the passport-less: an Affidavit of Identity in Lieu of Passport. It reads: "I wish to use this document in lieu of a passport which I, a state-less person, cannot obtain at present."[2]

> The passport is the noblest part of a human being. Nor does it come into the world in such a simple way as a human being. A human being can come about anywhere, in the most irresponsible manner and with no proper reason at all, but not a passport. That's why a passport will always be honoured, if it's a good one, whereas a person can be as good as you like, and still no one takes any notice.
>
> Bertolt Brecht, *Refugee Conversations* [1940] 2019: 8

Brecht's figurative image of the passport as a 'body-part' has been adopted, quite literally, as the migrant management vision for the Schengen Era. Once regarded as a hallmark of modernity, the passport is becoming, in this view, an obsolete relic that biometrics ought to replace. Since the 1980s, European institutions have founded pan-European databases with the shared purpose of storing a wide range of personal data on third-country nationals. The

Seit den 1980er Jahren haben europäische Institutionen paneuropäische Datenbanken mit dem gemeinsamen Ziel eingerichtet, ein breites Spektrum an personenbezogenen Daten von Drittstaatsangehörigen zu speichern. Die am weitesten verbreitete dieser Datenbanken und die erste, die mit biometrischen Identifikatoren experimentiert, ist Eurodac, in der Fingerabdrücke (und bald auch Gesichtsbilder) von Geflüchteten und ›irregulären Migrant:innen‹ gespeichert werden. Papiere und Passfotos werden von politischen Entscheidungsträger:innen mit wachsendem Misstrauen betrachtet. Sie können, so wird argumentiert, manipuliert, gefälscht oder zerstört werden. Wenn der Plan von Eurodac erfolgreich ist, wird die Biometrie den Körpern der Migrant:innen eine offizielle Identität einschreiben, zusätzlich zu der Last der europäischen Binnengrenzen, die jetzt hauptsächlich auf Nicht-EU-Migrant:innen lastet. Diese Strategie zielt darauf ab, Pässe als Vermittler zwischen Staaten, Organen und der Wahrheit zurückzudrängen und letztlich zu eliminieren. Da die EU die Mitgliedsstaaten unter Druck setzt, von allen Migrant:innen ohne Papieren Fingerabdrücke zu nehmen, werden Migrant:innen ohne Papiere umgekehrt zu den am meisten dokumentierten Subjekten im heutigen Europa. Dies wirft einige kritische Fragen auf: Sind wir wirklich Zeugen des Untergangs des Passes? Und wenn Pässe bisher im Mittelpunkt der Migration standen, wie würde das, was als ›verkörperte Pässe‹ oder ›der Körper als Pass‹ bezeichnet werden könnte, die Erfahrung von Zwangsmigration und Grenzübertritt verändern?

Als ich mich entschloss, eine Ethnografie über Eurodac zu schreiben, um die gelebte

most pervasive of these databases, and the first to experiment with biometric identifiers, is Eurodac, which stores fingerprints (and soon facial images) of refugees and 'irregular migrants'. Papers and passport photos are treated by policy makers with growing suspicion. They can, it is argued, be manipulated, faked or destroyed. If Eurodac's plan succeeds, biometrics will imprint legal identities on migrants' bodies, in addition to the burden of Europe's internal borders, which now falls principally on non-EU migrants. This strategy aims to relegate, and ultimately eliminate, passports as mediators between states, bodies and truth. As the EU pressures its member-states to fingerprint every undocumented migrant, undocumented migrants, conversely, become the most documented subjects in Europe today. This raises some critical questions: Are we really about to witness the demise of the passport? And if passports have been at the heart of migration until now, how would what we could call 'embodied passports' or the 'body-as-passport' transform the experience of forced migration and border-crossing?

When I set out to write an ethnography of Eurodac, to explore the lived reality of body-passports on the move, I expected to find what Eurodac has been promising: a migration-management apparatus through which 'truth' is extracted directly from migrants' bodies, rendering papers redundant. But this is hardly the case. Documents are still in use everywhere. To migrants, navigating one's paper-trail remains as crucial as navigating one's phys-

Realität von Körperpässen auf Reisen zu erforschen, erwartete ich, das zu finden, was Eurodac verspricht: einen Apparat zur Migrationssteuerung, durch den die ›Wahrheit‹ direkt aus den Körpern der Migrant:innen extrahiert wird und der Papiere überflüssig macht. Aber das ist kaum der Fall. Dokumente werden nach wie vor überall verwendet. Für Migrant:innen bleibt es genauso wichtig, den Weg ihrer Papiere zu navigieren wie ihre eigene körperliche Reise. Hält Eurodac also nicht, was es verspricht?

Gilles de Kerchove, ehemaliger EU-Koordinator für die Terrorismusbekämpfung, erläuterte 2017, warum er die biometrische Erfassung der europäischen Migrant:innen als so wichtiges Anliegen ansieht. »Unsere vordringliche Herausforderung besteht darin, Datenbankabfragen mit biometrischen Daten und insbesondere mit Gesichtsbildern zu ermöglichen«, sagte er. »Wir brauchen sie, weil viele Menschen, die nach Europa kommen, entweder ganz ohne Papiere oder mit falschen oder gestohlenen Papieren reisen. Biometrische Daten sind die einzige Möglichkeit, um zu erkennen, wer diese Menschen wirklich sind.«[3]

Kerchoves Aussage gibt einen Hinweis darauf, warum der verstärkte Einsatz biometrischer Daten nicht wirklich zu einer papierlosen Migration führt. Einerseits entspricht der Trend der Tatsache, dass Papiere mit wachsendem Misstrauen behandelt werden, weil sie gefälscht oder von einer anderen Person verwendet werden können. Andererseits erwecken Migrant:innen, die überhaupt keine Papiere oder keine als wichtig erachteten Dokumente besitzen, genau dadurch den Verdacht des Betrugs. Auch wenn eine solche vermeintliche Unvollständigkeit für sich genommen nicht ausreicht, um die Verdächtigen zu belasten,

ical journey. Is Eurodac then failing to deliver on its promise?

Gilles de Kerchove, a former EU counterterrorism coordinator, explained in 2017 why he viewed recording Europe's migrants biometrically as such a pressing concern. "Our urgent challenge is to feed and query our databases with biometrical data and especially facial imaging," he said. "We need it because many people entering Europe are travelling with no documents or false documents or stolen documents. Biometrical data is the only way to identify who these people really are."[3]

Kerchove's statement hints at why the increased use of biometrics does not actually lead to paperless migration. On the one hand, the trend resonates with the growing suspicion accorded to papers, because they can be falsified or used by a person/persons to whom they have not been issued. But conversely, if a migrant has no papers at all or doesn't possess those documents deemed essential, this itself arouses suspicions of fraud. Although such perceived incompleteness is not enough, in isolation, to incriminate those suspected, such official doubts are unquestionably a disadvantage to those negotiating the system. Moreover, applicants need paper documents to validate their biometric records. Such papers are also used to authenticate issues that are not visible in any straightforward sense, such as age, name, nationality or birthplace. The biometric encounter also eliminates some forms of print documents while generating others: whether a simple paper documenting biometric registrations or a

sind solche offiziellen Zweifel fraglos ein Nachteil für diejenigen, die mit dem System verhandeln. Außerdem werden Papiere benötigt, um die biometrischen Daten aussagekräftig zu machen. Sie werden verwendet, um Dinge zu authentifizieren, die nicht vollständig sichtbar sind, wie Alter, Name, Nationalität oder Geburtsort. Die biometrische Begegnung beseitigt einige Formen von gedruckten Dokumenten, während andere entstehen: ein einfaches Papier, das die biometrischen Registrierungen dokumentiert, oder ein biometrischer Reisepass, der den biometrischen Abdruck physisch speichert.

Viele Wissenschaftler:innen haben versucht, die Geschichte des Passes zu schreiben. Kurz zusammengefasst, der Pass ist in Europa während des Ersten Weltkrieges als Mittel zur Kontrolle der Bevölkerung erfunden wurde, im Gegensatz zur freien Mobilität. Der österreichische Schriftsteller Stefan Zweig berichtete, dass er vor dem Krieg nach Indien und nach Amerika reisen konnte, »ohne einen Paß zu besitzen oder überhaupt je einen gesehen zu haben.« (1965 [1942]: 371–372). Der Pass entstand während des Krieges als vorübergehende Notmaßnahme, um Menschen daran zu hindern zu fliehen und der Einberufung zu entgehen. Was als Ausnahme in Kriegszeiten gedacht war, wurde in Friedenszeiten zur Regel. Doch um eine umfassende Geschichte – oder umfassende Geschichten – des Passes zu schreiben, muss zunächst ermittelt werden, was ein Pass überhaupt ist. Als Objekt enthält er mehrere sich überschneidende und regulierende Technologien: ein Zeugnis der Staatsbürgerschaft, ein Laissez-passer (Reisedokument) und ein Passfoto. Wie bei jeder Technologie des Herrschen ist eine Reihe von

biometric passport that physically stores the biometric print.

Many scholars have tried to write the history of the passport. The simple story is that the passport was invented in Europe during World War I as a means of enforcing population containment, as opposed to population movement. The Austrian author Stefan Zweig wrote that he could travel to India and to America before the war "without a passport and without ever having seen one" (Zweig [1942] 2013: 410). The passport emerged during the war as a temporary, emergency measure to prevent people from fleeing and avoiding conscription. What was conceived of as a wartime exception became the rule in times of peace. But to write a comprehensive history, or histories, of the passport, one must first ascertain what a passport is. As an object, it contains several overlapping and regulating technologies: a testament of citizenship, a *laissez-passer* (travel document) and a passport photo. As with every technology of governance, a range of conditions are required for it to emerge in the first place. An international nation-state system, a standardized global network of borders and movement surveillance and a regime of visual regulation are principal prerequisites. Beyond merely representing these historical conditions, the passport produces the fantasies that sustain them as 'real'.

The American philosopher Susan Buck-Morss writes that the passport promotes the fantasy that:

Bedingungen erforderlich, damit sie überhaupt erst entstehen kann. Ein internationales nationalstaatliches System, ein standardisiertes globales Netzwerk von Grenzen und Bewegungsüberwachung sowie ein Regime der visuellen Regulierung sind die wichtigsten Voraussetzungen dafür. Über die bloße Darstellung dieser historischen Bedingungen hinaus produziert der Pass die Fantasien, die die Bedingungen als ›real‹ aufrechterhalten.

Die amerikanische Philosophin Susan Buck-Morss schreibt, dass der Pass die Vorstellung fördert, dass

> Staatsgrenzen etwas Substanzielles sind, dass sie wirklich existieren – dass ein bestimmter Staatsapparat einen Teil der Welt ›besitzt‹, so wie ein Privatmann sein Haus ›besitzt‹, ein Kapitalist sein Geschäft, ein Bauer sein Feld, ein Mensch seinen Körper – nur dass der Staat all diese Dinge zuerst besaß. (1993: 66–77)

So tragen Pässe dazu bei, die kollektive Imagination zu erzeugen, die die staatliche Souveränität erfordert. Doch was geschieht mit dem Pass und seiner Scheinmacht in einer Welt, in der die Ungleichheit des globalen Wohlstands eskaliert? In einer Welt, in der Menschen gerade deshalb fliehen, weil sie von ihrem Staat verleugnet werden? Oder aber, weil ihr Staat so enteignet ist, dass er den Bürger:innen, die ihm ›gehören‹, keinen souveränen Schutz bieten kann? In solchen Fällen wird der Pass selbst zu einem Ort der globalen Enteignung, da die Migrant:innen erkennen, wie ihre Pässe zu entwerteten Artefakten geworden sind, die weder die Vorteile der Staatsbürgerschaft noch die Vorteile eines Reisedokuments absichern.

State boundaries were substantive, that they really existed – that a particular state apparatus 'owned' a part of the world, in the same way as a private citizen 'owned' his home, a capitalist his business, a farmer his field, a person his or her own body – except that the state owned all of these first. (1993: 66–77)

Thus, passports contribute to producing the collective imagination that state sovereignty requires. But what happens to the passport and its make-believe power in a world of escalating global-wealth inequalities? A world in which people flee precisely because they have been disowned by their state? Or, alternatively, because their state is so dispossessed that it cannot provide sovereign protection to the citizens that it 'owns'. In such cases, the passport itself becomes a site of global dispossession, as migrants realize how their passports have become devalued artefacts that fail to secure either the benefits of citizenship or the advantages of *laissez-passer* movement.

And if the passport (re)produces relations of ownership, can this transaction work in reverse? Can someone buy into citizenship by purchasing a passport? This question preoccupies both refugees and the global elite. The German Government website *Rumours about Germany: Facts for Migrants* attempts to set the record straight. Under the heading "What is a passport and why can I not buy one?",[4] it purports to "educate" migrants that passports are state property, and thus "cannot

Wenn der Pass Eigentumsverhältnisse (wieder) herstellt, kann dann diese Transaktion auch umgekehrt funktionieren? Kann sich jemand durch den Kauf eines Passes in die Staatsbürgerschaft einkaufen? Diese Frage beschäftigt sowohl die Geflüchteten als auch die globalen Eliten. Die Webseite der deutschen Regierung *Rumours about Germany: Facts for Migrants* (Gerüchte über Deutschland: Fakten für Migrant:innen) versucht, die Dinge richtigzustellen. Unter der Überschrift »What is a passport and why can I not buy one?« (Was ist ein Pass und warum kann ich keinen Pass kaufen?)[4] möchte die Seite Migrant:innen darüber »aufklären«, dass Pässe Staatseigentum sind und daher »nicht gekauft werden können«. »Jeder Pass, der im Internet oder in einzelnen Ländern zum Kauf angeboten wird, ist daher entweder gestohlen oder eine kriminelle Fälschung«, heißt es dort. Zusammen mit den Fakten über die käufliche Staatsbürgerschaft liest sich dies wie eine unverfrorene Lüge. Viele Länder, darunter die EU-Mitgliedsstaaten Österreich, Zypern und Malta, bieten Pass-durch-Investment-Programme an. Firmen wie Henley and Partners werben für Pass-durch-Investment-Programme in Flugzeitschriften und mit Internetanzeigen. Laut der Website von Henley können Kunden heute für eine »Mindesteinlage« von drei Millionen Euro und »ohne vorherige Wohnsitzanforderungen« einen Reisepass für den EU-Mitgliedsstaat Österreich erhalten.[5]

---

be bought". "Any passport offered for sale on the Internet or in individual countries is therefore either stolen or a criminal forgery," it states. Laid side-by-side with the facts of purchasable citizenship, this reads like a barefaced lie. Many countries, including EU members Austria, Cyprus and Malta, offer passport-by-investment schemes. Firms like Henley and Partners advertise passport-by-investment programs in in-flight magazines and by using internet ads. According to Henley's website, clients can obtain a passport for the EU member-state Austria today for a "minimum contribution" of three million euros and "without prior residence requirements".[5]

The Paper Trail of Migration

Chios Hotspot, Griechenland, 2018

Während meiner Recherchen auf Chios erhielt ich auf meine Frage »Gibt es Neuigkeiten?« fast immer die gleiche Antwort: »Keine Neuigkeiten, ich warte nur auf meine *kharti*« (χαρτί, griechisch für ›Papier‹). *Kharti* bedeutete für die verschiedenen Menschen, mit denen ich sprach, unterschiedliche Dinge: Einige warteten auf ihre Erstregistrierungsunterlagen, die für alles von der Beantragung von Asyl bis zur medizinischen Versorgung im Hotspot entscheidend sind. Andere warteten auf das ›blaue Papier‹, mit dem die Beschränkungen der Freizügigkeit innerhalb Griechenlands aufgehoben wurden, so dass die Migrant:innen den Hotspot in Richtung Festland verlassen konnten. Die wenigen Glücklichen waren diejenigen, die nach einer erfolgreichen Asylanhörung – auf die die Geflüchteten bis heute monatelang unter erbärmlichen Bedingungen warteten – das ›schwarze Papier‹ erhielten, eine Bescheinigung, die den Inhaber:innen mitteilt, dass sie auf den ›Reiseausweis für Flüchtlinge‹ warten müssen, der auf Chios umgangssprachlich »der Pass« bezeichnet wird.

An einem Sommerabend veranstaltete Firas eine ›Passparty‹. Wir lernten uns kennen, als ich den 25-jährigen Friseur und Model aus dem Irak eines Tages vom Lager in die Stadt mitnahm. In den vorangegangenen achtzehn Monaten hatte er in einem Metallcontainer im Lager Vial gelebt, das berüchtigt ist und stark kritisiert wird. Er hatte gerade sein ›schwarzes Papier‹ erhalten, und sobald er den so genannten Pass hatte, wollte er den

Chios Hotspot, Greece, 2018

During my research in Chios, asking an asylum seeker "Any news?" would almost always elicit the same response: "No news, just waiting for my *kharti*" (χαρτί, Greek for 'paper'). *Kharti* meant different things to different people I spoke with: Some were waiting for their initial registration paper, crucial for everything from claiming asylum to receiving healthcare at the hotspot. Others waited for the 'blue paper' that lifted freedom-of-movement restrictions imposed within Greece, allowing migrants to leave the hotspot for the mainland. The lucky few were those who, after a successful asylum-application interview – for which refugees waited and still wait months in squalid conditions – were issued with the 'black paper', a certificate that tells holders to wait for the 'refugee travel document', known colloquially on Chios as 'the passport'.

One summer evening, Firas was throwing a 'passport party'. He was a 25-year-old barber and model from Iraq I had met when I gave him a ride from the camp to the city one day. During the previous 18 months, he had been living in a metal container in Vial Camp, which has a notorious reputation and has been heavily criticised. He had just received his 'black paper', and once he got the so-called 'passport' he was planning to leave the hotspot to start a new life on the mainland.

His party was scheduled for 10 p.m. on 10 July 2018, just after the World Cup semi-final in which France beat Belgium. The

Hotspot verlassen, um ein neues Leben auf dem Festland zu beginnen.

Seine Party war für 22.00 Uhr am 10. Juli 2018 geplant, kurz nach dem Halbfinale der Fußballweltmeisterschaft, in dem Frankreich Belgien besiegt hat. Als Ort wählte er den Standardort für Passpartys: das äußerste Ende der Hafenmole unter einem Leuchtturm. Die Ecke des Leuchtturms, die sich tief ins Meer erstreckt, ist ein intimer und zugleich festlicher Ort. Mit Blick vom Meer auf die belebte Promenade fühlen sich die Gäste mitten in der Stadt und gleichzeitig unsichtbar. Mitten im Geschehen und doch abgeschieden, können die Geflüchteten laut musizieren, tanzen, trinken und lachen, ohne bemerkt zu werden. Sie sind abgeschirmt vom Rauschen der Wellen, die sie überlebt haben, um hierher zu gelangen, und von denen sie träumen und vor denen sie sich fürchten: beunruhigende Visionen, wie sie eines Tages ihre Reise fortsetzen werden.

Während auf der Tanzfläche Songs von Nicki Minaj und Shakira ertönten, gelang es mir, mit Amir zu sprechen, der in der Woche zuvor seine Passparty gefeiert hatte. »Das ändert alles«, sagte er und starrte auf das dunkle Meer. »Es verändert das Meer, es verändert die Straße, es verändert meine Geschichte.« Es gebe einer Reihe von Ereignissen, Erinnerungen und Entscheidungen einen Sinn und ein Ziel, von denen er jetzt – im Nachhinein – sagen könne, dass sie nicht umsonst waren. Es ändere sich so viel, sagte er, dass ich ihn vielleicht noch einmal befragen müsse. Nicht, weil er mir neue Informationen zu erzählen hätte, sondern weil, wie er erklärte, der Pass ihm die Möglichkeit gebe, über seine Reise auf eine ganz andere Weise nachzudenken.

Aber Yasir, ein Politikwissenschaftler aus

location he chose was the default place for passport parties: the farthermost end of the port's pier, under a lighthouse. Stretching deep into the sea, the lighthouse corner is a place both intimate and festive. Overlooking the bustling promenade from the sea, guests feel like they are in the middle of town and invisible at the same time. Both central and removed, refugees can play loud music, dance, drink and laugh without being noticed. They are shielded by the sound of the waves, the very waves they survived to get here, and which they dream about and dread: fraught visions of one day sailing onwards on their journeys.

As Nicki Minaj and Shakira songs blast from the dancefloor, I manage to have a word with Amir, who had celebrated his passport party the previous week. "This changes everything," he says, staring at the dark sea. "It changes the sea, it changes the road, it changes my story." It gives meaning, and purpose, to a series of events, memories and decisions that now – in hindsight – he can say were not in vain. It changes so much, he says, that I might need to interview him again. Not because he has new information to tell me, but because, he maintains, the passport will give him the ability to reflect on his journey in an entirely different way.

But Yasir, a political scientist from Somalia who was still waiting for his documents, tells me that refugee papers do not necessarily provide a conclusion to a journey of migration. He says it can throw some individuals from a "waiting-in-uncertainty" mode back into flight mode. "When we get the papers," he says, "is when we

Somalia, der immer noch auf seine Papiere wartete, sagte mir, dass Flüchtlingspapiere nicht unbedingt den Abschluss einer Migrationsreise bedeuten würden. Er meinte, dass sie manche Menschen von einem »Warten-in-Ungewissheit«-Modus zurück in den Fluchtmodus versetzen könnten. »Wenn wir die Papiere bekommen«, sagte er, »fangen wir an, uns wirklich Sorgen zu machen. Nach Monaten des Wartens fühlen wir uns wieder unsicher«.

Ein Geflüchteter weiß aus Erfahrung um die Vergänglichkeit, Unstetigkeit und Unbeständigkeit von Dokumenten und ist sich bewusst, dass offizielle internationale Schutzmaßnahmen fadenscheinige Dinge sind, wenn sie auf die Probe gestellt werden. Papiere können ein gewisses Maß an Schutz bieten und die Bewegungsfreiheit erleichtern. Da die Asylpolitik der einzelnen Staaten jedoch sehr unterschiedlich ist, stellen Papiere auch ein Dilemma dar: Entscheidet man sich dafür, sich auf ein bestimmtes Dokumentensystem zu verlassen, oder entscheidet man sich dafür, ein anderes zu verwenden? Während meiner Forschung stellte ich fest, dass die meisten Asylbewerber:innen monatelang darauf warteten, das blaue *kharti* zu erhalten und einen Termin für eine Anhörung im nächsten Jahr zu bekommen. Während dieser Zeit durften sie nicht arbeiten und sind auf eine monatliche Unterstützung von neunzig Euro durch den UNHCR (United Nations High Commissioner for Refugees) angewiesen. Viele Geflüchtete interpretierten das extrem langwierige Anhörungsverfahren als ein informelles Signal der griechischen Regierung an die Geflüchteten, dass sie in einen anderen EU-Staat ziehen sollten, eine indirekte Art zu sagen, dass das griechische System überlastet ist. Aber selbst

really start to worry. We feel insecure again after months of waiting."

A refugee knows about the transient, fickle and unstable nature of documents from experience and is conscious that official international protections are flimsy things if put to the test. Papers may offer some degree of protection, as well as facilitating movement. But because states differ vastly in their asylum policies, papers also pose a dilemma: Do you decide to rely on one specific documentation regime or do you opt to move to another? During my research, most asylum-seekers waited for months just to receive the blue *kharti* and an asylum interview date scheduled for the following year. They are not permitted to work during this period and are forced to rely on a 90-Euro a month allowance from the UNHCR. Many refugees interpreted the extremely protracted interview procedure as an informal signal from the Greek government to refugees that they should move to another EU state, an indirect way of saying that the Greek system was overloaded. But even for the lucky few who receive formal refugee protection, in Greece or elsewhere, this status is often only temporary.[6] This impossible dilemma is why, for Yasir, being issued with a paper is a thoroughly ambivalent blessing, when all the uncertainties that accompany this document are weighed up. He reads it as a call to flee again: "We have to prepare for another trip. To take risks. It's like the hotspot was all a dream. You wake up, and you're back in İzmir before you have boarded the boat, facing the dangerous sea."

Abb. 1: Eine Passparty. Hafen von Chios, 2018. ©Romm Lewkowicz

Image 1: A passport party. Chios Port, 2018. ©Romm Lewkowicz

für die wenigen Glücklichen, die in Griechenland oder anderswo formellen Flüchtlingsschutz erhalten, ist dieser Status oft nur vorübergehend.⁶ Dieses unerträgliche Dilemma ist der Grund, warum die Ausstellung eines Papiers für Yasir ein durch und durch ambivalenter Segen war, wenn all die Ungewissheiten abgewogen wurden, die mit diesem Dokument einhergehen. Er verstand es als Aufforderung, erneut zu fliehen: »Wir müssen uns auf eine weitere Reise vorbereiten. Weitere Risiken eingehen. Es ist, als wäre der Hotspot nur ein Traum gewesen. Du wachst auf und bist wieder in İzmir, noch bevor man das Boot bestiegen hat und sich dem gefährlichen Meer stellt.«

Literatur | References

Agamben, Giorgio. 1998. *Homo Sacer: Sovereign Power and Bare Life*. Translated by Daniel Heller-Roazen. Stanford, CA.

Arendt, Hannah. 1951. *The Origins of Totalitarianism*. New York.

Arendt, Hannah. 1955. *Elemente und Ursprünge Totaler Herrschaft*. Frankfurt a.M..

Brecht, Bertold. 2019. *Bertold Brecht's Refugee Conversations*. Translated by Romy Fursland, Edited and Introduced by Tom Kuhn. London.

Brecht, Bertold. 1961. *Flüchtlingsgespräche*. Frankfurt.

Buck-Morss, Susan. 1993. Passports. *Documents*, 1, 66–77.

Zweig, Stefan. 1965. *Die Welt von Gestern: Erinnerungen eines Europäers*. Berlin.

Zweig, Stefan. 2013. *The World of Yesterday*. Translated by Anthea Bell. Lexington.

Anmerkungen

1. Alle Namen und einige Details wurden geändert, um die Privatsphäre der Personen zu schützen.
2. Dieses Dokument wird in der Library of Congress aufbewahrt und wurde digitalisiert. Siehe: Jerome Kohn: »World of Hannah Arendt«, *Library of Congress*, 11. November 2021, https://www.loc.gov/collections/hannah-arendt-papers/articles-and-essays/world-of-hannah-arendt/
3. Frédéric Simon: EU anti-terror czar: 'The threat is coming from inside Europe', *Euractive*, 22. März 2017, https://www.euractiv.com/section/freedom-of-thought/interview/eu-anti-terror-czar-the-threat-is-coming-from-inside-europe/
4. »What is a passport and why can I not buy one?«, *Rumours about Germany: Facts for Migrants*, abgerufen am 27. November 2018, https://rumoursaboutgermany.info/facts/what-is-a-passport-and-why-can-i-not-buy-one/
5. Henley & Partners, »The Global Leader in Residence and Citizenship by Investment«, Henley & Partners, abgerufen am 16. Oktober 2021, https://www.henleyglobal.com/
6. Um Anspruch auf internationalen Schutz nach der Flüchtlingskonvention von 1951 zu haben, müssen Antragsteller:innen eine begründete Furcht vor persönlicher Verfolgung durch einen Staat nachweisen. Dies bedeutet, dass vielen Geflüchteten, die vor den Bürgerkriegen in Syrien oder im Jemen fliehen, subsidiärer Schutz gewährt wird, im Gegensatz zum Schutz gemäß Konvention. Ersterer ist ein Status, den ein Staat jederzeit widerrufen kann, sollte er entscheiden, dass sich die Umstände, die den ursprünglichen Schutz rechtfertigen, geändert haben.

Notes

1. All names and some identifying details have been changed to protect the individuals' privacy.
2. This document is archived in the Library of Congress and has been digitized. See Jerome Kohn: "World of Hannah Arendt", *Library of Congress*, accessed 11 November 2021, https://www.loc.gov/collections/hannah-arendt-papers/articles-and-essays/world-of-hannah-arendt/
3. Frédéric Simon: EU anti-terror czar: 'The threat is coming from inside Europe', *Euractive*, 22 March, 2017, https://www.euractiv.com/section/freedom-of-thought/interview/eu-anti-terror-czar-the-threat-is-coming-from-inside-europe/.
4. "What is a passport and why can I not buy one?", *Rumours about Germany: Facts for Migrants*, accessed 27 November 2018, https://rumoursaboutgermany.info/facts/what-is-a-passport-and-why-can-i-not-buy-one/
5. Henley & Partners, "The Global Leader in Residence and Citizenship by Investment", Henley & Partners, accessed 16 October, 2021, https://www.henleyglobal.com/
6. To be eligible for international protection under the Refugee Convention of 1951, applicants must prove a well-founded fear of personal persecution by a state. This means that many refugees fleeing the civil wars in Syria or in Yemen are granted subsidiary protection, as opposed to convention protection. The former is a status a state can revoke at any time, should it rule that the circumstances justifying the initial protection have changed.

Abb. 2: Saul Steinberg, Pass, 1953, Mischtechnik auf Papier, 54,6×43,2 cm, Sammlung von Leon und Michaela Constantiner, ©The Saul Steinberg Foundation/Artists Rights Society (ARS), New York

Image 2: Saul Steinberg, Passport, 1953, Mixed media on paper, 54,6 × 43,2 cm, Collection of Leon and Michaela Constantiner, ©The Saul Steinberg Foundation/Artists Rights Society (ARS), New York

Saul Steinberg, The Passport
Leonie Karwath

Im Werk des Zeichners Saul Steinberg (1914–1999) ist der Pass ein wiederkehrendes Motiv mit starken biografischen Bezügen. Steinberg wird 1914 in der Nähe von Bukarest in Rumänien als Sohn jüdischer Eltern geboren. Während seines Architekturstudiums in Mailand beginnt er humoristische Zeichnungen zu veröffentlichen, erhält jedoch bald vom faschistischen Italien Berufsverbot. 1941 versucht er erstmals mit einem abgelaufenen Pass mit einem von ihm selbst gefälschten Stemmpel über Portugal in die USA zu fliehen. Da aber die Aufnahmequote für Rumänen bereits erschöpft ist, wird er von Ellis Island in die Dominikanische Republik abgeschoben. Von dort aus sendet er der Zeitschrift *The New Yorker* einige seiner Cartoons zu, was ihm schließlich die Türen für eine Einreise in die USA öffnet. Seine Zeichnungen für den *New Yorker* werden weltbekannt.

Saul Steinberg, The Passport
Leonie Karwath

The passport, with its strong biographical connotations, recurs repeatedly in the work of the cartoonist and illustrator Saul Steinberg (1914–1919). Born as the son of Jewish parents near Bucharest in Rumania in 1914, Steinberg began publishing humorous drawings while he was studying architecture in Milan. After receiving a work-ban from the fascist authorities in Italy, he attempted to flee via Portugal to the USA in 1941, using an out-of-date passport with a stamp he had falsified himself. But because the immigrant quota for Rumanians for that year was already exhausted, the US authorities deported him following his arrival at Ellis Island, to the Dominican Republic. From there he sent a number of cartoons to *The New Yorker*, which ultimately opened the doors necessary for him to enter the USA. His drawings for *The New Yorker* became famous across the globe.

Abb. 3: Passbild, 1951, Daumenabdrücke auf Papier, ©The Saul Steinberg Foundation/Artists Rights Society (ARS), New York

Image 3: Passport Photo, 1951, Thumbprints on paper, ©The Saul Steinberg Foundation/Artists Rights Society (ARS), New York

Abb. 4: Gruppenfoto, 1953, Tinte, Daumenabdrücke und Gummistempel auf Papier, 35,6×27,9 cm, Sammlung von Richard und Ronay Menschel, ©The Saul Steinberg Foundation/Artists Rights Society (ARS), New York

Image 4: Group Photo, 1953, Ink, thumbprints, and rubber stamp on paper, 35.6 × 27.9 cm, Collection of Richard and Ronay Menschel, ©The Saul Steinberg Foundation/Artists Rights Society (ARS), New York

The Paper Trail of Migration

Abb. 5: Fingerabdruck-Landschaft, 1950, Fingerabdrücke und Bleistift auf Papier, 37×28,3 cm, Beinecke Rare Book and Manuscript Library, Yale University, ©The Saul Steinberg Foundation/Artists Rights Society (ARS), New York

Image 5: Fingerprint Landscape, 1950, Fingerprints and pencil on paper, 37 × 28.3 cm, Beinecke Rare Book and Manuscript Library, Yale University, ©The Saul Steinberg Foundation/Artists Rights Society (ARS), New York

Abb. 6: Unbenannt, ca. 1950–53, Fingerabdrücke auf Papier, ©The Saul Steinberg Foundation/Artists Rights Society (ARS), New York

Image 6: Untitled, ca. 1950–53, Fingerprints on paper, ©The Saul Steinberg Foundation/Artists Rights Society (ARS), New York

Die Papierspur der Migration

Von Puppen, Bären, Drachen und Waffen
Spielzeug, Flucht und Migration

On Dolls, Bears, Kites and Weapons
Toys, Flight and Migration

Peter J. Bräunlein

Abb. 1: Während der Adventszeit 2015 beschenkt die (damalige) Familienministerin Manuela Schwesig Flüchtlingskinder mit Plüschtieren bei ihrem Besuch auf der Berlin Messe. Wer für die Auswahl des verschenkten Spielzeugs und der zu beschenkenden Kinder sorgte, ist unbekannt. Unklar bleibt auch, wie sich die Kinder mit den neuen Plüschtieren fühlen. Hätten sie lieber was anderes gehabt? Welche Namen gaben sie ihren neuen Gefährten? ©Fabrizio Bensch/Reuters

Image 1: Manuela Schwesig, then Federal Family Minister, giving refugee children presents of cuddly toys at the Berlin Trade Fair during Advent 2015. We don't know who selected the toys that were to be gifted or the children who were to receive presents. It also remains unclear how the children in the photo feel about their new cuddly toys. Would they have preferred something else? What names did they give their new companions? ©Fabrizio Bensch/Reuters

»Spielen tut gut«: Unter dieser Überschrift wirbt die Hilfsorganisation World Vision um Spielzeugspenden. Auf ihrer Webseite ist zu lesen:

"Play is good for us" ["Spielen tut gut"] – this is the slogan used by aid organisation World Vision in their German-language campaigns to advertise for toy donations. They state on their website:

Kinder leiden ganz besonders unter der Flucht. Im Libanon gibt es viele syrische Flüchtlingskinder, die alles [zurücklassen] mussten und nur wenig sinnvolle Beschäftigungsmöglichkeiten haben. Um ihnen ein Stück Kindheit [zurückzugeben] und etwas Freude in den oft tristen Alltag zu bringen, wollen wir so viele Kinder wie möglich mit Spielsachen versorgen. Mit deiner Spende schenkst du altersgerechtes und pädagogisch wertvolles Spielzeug wie Bücher, Lernpuzzle oder Mal-Utensilien. Diese Sets helfen den Kindern auch dabei, ihre häufig traumatischen Fluchterfahrungen spielerisch zu verarbeiten.[1]

Für 48 Euro kann ein Spielzeugset erworben werden, das an bedürftige Kinder versandt wird. In einer Holzkiste enthalten sind, wie auf der Abbildung der Webseite erkennbar, Teddybär und Stoffpuppe, Bauklötze mit lateinischen Buchstaben, ein Puzzle, ein roter Ball, Buntstifte sowie ein kleines Holzauto. Unklar bleibt, ob die abgebildeten Spielsachen nur als Beispiel dienen oder tatsächlich auch in dem zu erwerbenden Set enthalten sind.

Explizit adressiert der Spielzeugspendenaufruf die Gefühle der zu beschenkenden Kinder: Leid, Freude, trister Alltag und traumatische Fluchterfahrung. Die Gefühle potenziell Spendender – Hilflosigkeit und Rührung – werden implizit adressiert. Gleichzeitig wird die Funktion von Spielzeug thematisiert: Kinder sollen mit Spielzeug »versorgt« werden, und zwar mit »altersgerechtem und pädagogisch wertvollem« Spielzeug. Das Spielzeug soll »ein Stück Kindheit [zurückgeben]«,

When people have to flee, it is children who suffer the most. There are many Syrian refugee children in Lebanon, who had to leave everything behind and who now have only few meaningful activities they can engage in. To give them back a piece of their childhood and to bring a little joy into their often-bleak everyday lives, we want to provide as many of these children as possible with things to play with. You can use your donation to gift age-appropriate and educationally valuable toys, including books, learning puzzles, or painting and drawing utensils. These sets also help the children to work through their often traumatic experiences of flight.[1]

One of these toy sets can be purchased for €48 to then be sent to needy children. Donors looking at the photos on the website can discern a wooden box containing a teddy bear and a cloth-doll, building blocks with Latin alphabet letters on them, a puzzle, a red ball, felt-tip pens and a little wooden car. It remains unclear whether the toys presented are just meant to serve as examples or whether they're actually part of the purchasable set.

The appeal for toy donations is directed explicitly towards the feelings of the kids receiving the presents: sorrow, joy, bleak everyday life and traumatic experience of flight. The feelings of potential donors, which may include both helplessness and a sense of being touched at the same time, are addressed more implicitly. Parallel to this, the appeal focuses on the

»Freude in den oft tristen Alltag« bringen und helfen, die »häufig traumatischen Fluchterfahrungen spielerisch zu verarbeiten«.

Dieser erwachsene Blick auf Spielsachen ist erzieherisch motiviert. Spielzeug gehört demnach zur menschlichen Grundversorgung, und Spielen tut gut, und zwar dann, wenn Kinder *richtig* spielen, d.h. altersgerecht und pädagogisch wertvoll. Aus dem Spiel kann dann sogar sinnvolle Arbeit werden, nämlich Trauma-Arbeit. Im Spielzeug treffen sich, so betrachtet, Gefühle, pädagogische Absicht und Annahmen, was Kindsein bedeutet. Diese Sichtweise ist in westlichen Gesellschaften weit verbreitet und dort nahezu unvermeidlich. Das hat historische und kulturelle Gründe.

Ab dem 17. Jahrhundert begann in den westlichen Gesellschaften die ›Entdeckung der Kindheit‹ als eine eigene Phase des Menschseins. Diese Vorstellung wurde im Verlauf des 18. und 19. Jahrhunderts ausdifferenziert (Ariès 1975). Rousseau forderte den Schutz der Kindheit und die Freiheit zur Selbstentfaltung. Anders als im Mittelalter galten Kinder nun als reine Seelen, unschuldig, ausgestattet mit einer eigenen Gefühlswelt und als formbar. Unterschiedliche Spielaktivitäten wie Bewegungsspiel, Gemeinschaftsspiel, Verkleiden, Märchenerzählen oder eben auch das Spiel mit eigens dafür angefertigten Artefakten, vorwiegend Puppen, sind erst seit 300 Jahren ausschließlich Kindern vorbehalten. Die Bedeutung von handwerklich hergestelltem Spielzeug erlebte in der Biedermeierzeit cinen besonderen Schub, insbesondere durch die Entwicklung des Weihnachtsfestes. Dieses ritualisierte Bescheren der Kinder mit Spielzeug entstand

function of the toys: Children should be "provided for" with toys, and, more concretely, with "age-appropriate and educationally valuable" toys. The intention is that the toys "give [the children] a piece of their childhood back", while bringing "a little joy into their often bleak daily lives", and helping the children "to work through their often traumatic experiences of flight".

This adult perspective on toys is motivated by an attitude that has emerged from debates on how to bring up children. In this view, toys are part of basic human basic needs, and play is good for us, though this latter claim is made dependent on the condition that children play *in the right way*, i.e., in an age-appropriate and educationally valuable manner. Meaningful labour can even emerge from this play, namely, the task of dealing with trauma. Seen in this way, feelings, educational intentions and assumptions about what being a child means all meet in toys as theorized objects. This perspective is so widely spread in Western societies that it appears in these places as something almost unavoidable. The historical and cultural reasons for this are telling.

The 'discovery of childhood' as a distinctive phase of being human began in Western societies in the 17th century. What this concept of childhood meant to people became more nuanced during the course of the 18th and 19th centuries (Ariès 1975). Famously, Rousseau demanded that childhood be protected, and that children be allowed the freedom to develop autonomously. In contrast to perceptions in the Middle Ages, children were now seen as pure souls, innocent, with a world of feelings at their disposal

in eben dieser Zeit als zunächst großbürgerliches Brauchtum, das nach und nach von allen anderen Gesellschaftskreisen adaptiert wurde.

Wenn syrischen Kindern in einem libanesischen Flüchtlingslager mit Buchstabenbauklötzchen, Buntstiften und einem Puzzle Freude geschenkt und ein Stück Kindheit zurückgegeben werden soll, dann schwingt hier die bürgerliche Mentalitätsgeschichte Europas mit. Die erzieherischen Auffassungen, die die bürgerliche Entdeckung der Kindheit seit der Aufklärung und der Romantik nach sich zogen, wirken sich nachhaltig auf die Vorstellung von Spiel und Spielzeug aus. Vor diesem Hintergrund wird Spielzeug mitunter auch zum Politikum. Dies gilt beispielsweise für Spielzeugaktionen von deutschen Politiker:innen, die vorwiegend zur Weihnachtszeit medienwirksam in Flüchtlingsunterkünften inszeniert werden. Teddybären, Puppen, Bälle und Plüschtiere zaubern nicht nur Freude in Kindergesichter, die von Pressefotografen abgelichtet werden, sondern signalisieren den potenziellen Wählerinnen und Wählern: Seht her! Wir zeigen Mitgefühl und kümmern uns um die Opfer von Krieg und Vertreibung.

Klar dürfte sein, dass flüchtende oder geflüchtete Kinder im Libanon und andernorts auch ohne Spielzeugspenden spielen. Denn beim Spielen, da ist sich die Forschung einig, handelt es sich um ein Verhalten, das in der Natur von Menschen (und Tieren) liegt (Eigen und Winkler 1985). Ohne Spiel, so erklärt uns der Historiker Johan Huizinga, gebe es keine Kultur (Huizinga 1938). Verhaltensbiologisch betrachtet, dienen Körperspiele, Rol-

and also mouldable. Distinctive activities of play including games involving active movement, collaborative games, dressing up, narrating fairy-tales, or simply playing with artefacts created specifically for that activity – principally dolls – have been the exclusive preserve of children only for the last 300 years or so. But handmade toys for children only really came to prominence during the 19th century's Biedermeier era and particularly through the development of Christmas festivities. This was the period in which the ritualized giving of toys as presents to children established itself. Limited at first to the *haute bourgeoisie*, the tradition gradually established itself in all strata of society.

When Syrian kids in a Lebanese refugee camp are meant to receive a little joy and a piece of their childhoods back – through playing with lettered building blocks, felt-tip pens and a puzzle –, then we know that this history of bourgeois mentalities in Europe has its finger in the pie. The attitudes to upbringing that came with the Enlightenment and Romanticism have had a lasting impact on people's notions of what play is and what toys are. Against this backdrop, toys themselves become a political football. This rings especially true for toy campaigns carried out by German politicians, which are staged principally at Christmas time in refugee hostels and camps. Teddy bears, dolls, balls and cuddly toys are meant not merely to draw a look of joy in children's faces, which are then snapped up by press photographers; they also signal to potential voters: Look at this! We're able to show empathy

lenspiele und Nachahmungsspiele dem Einüben von körperlicher Geschicklichkeit und Sozialverhalten. Das Spiel mit Dingen, sei es konstruktiv oder destruktiv, dient dazu, Herrschaft über diese Dinge zu erlangen. Stein, Feder, Ast, Sand, Glasflasche, Blechdose und Plastiktüte – welche Sachen zum Spielen dienen, ist letztlich beliebig. Handwerklich oder industriell hergestelltes Spielzeug ist somit nur eine Sonderform von Spielzeug, die, wie erwähnt, eng mit der Entwicklung bürgerlicher Vorstellungen von Kindsein und Familie verbunden ist und die ab dem 19. Jahrhundert einen eigenen Produktionszweig hervorbringt. Das Zentrum dieser Spielzeugindustrie war zunächst Deutschland, verlagerte sich dann in die USA, bis in der zweiten Hälfte des 20. Jahrhunderts vor allem China marktbeherrschend wurde.

Das von außen betrachtet reichlich sinnlose Treiben des kindlichen Spiels erfüllt einen tieferen, letztlich überlebenswichtigen Sinn, gesteuert von Neugierde, Lernbereitschaft und Fantasie. Motiviert und belohnt wird diese Aktivität durch Spaßempfinden und ›Flow‹-Erleben, also dem Gefühl des Aufgehens in einer Tätigkeit (Csíkszentmihályi 2000).

Diese instrumentelle Sichtweise der Verhaltensforschung und Pädagogik kontrastiert mit einer philosophischen Wertschätzung des Spiels. Hier wird nicht das entwicklungspsychologisch Funktionelle des Spiels hervorgehoben, sondern sein utopisches Potenzial. »Der Mensch spielt nur, wo er in voller Bedeutung des Worts Mensch ist, und er ist nur da ganz Mensch, wo er spielt« (Schiller 1795: 88). Mit dieser berühmt gewordenen Sentenz feierte Friedrich Schiller den *Homo*

and look after the victims of war and persecution!

It should be obvious that children will play anyway, both while fleeing and while living as refugee children in Lebanon and elsewhere, regardless of whether they have donated toys. Researchers agree that play is a form of behaviour inherent to both human and animal natures (Eigen and Winkler 1985). The historian Johan Huizinga elucidates how without play there would be no culture (1938). Playing with things, whether this play is constructive or destructive, serves the purpose of achieving mastery over these things. Stones, feathers, branches, glass bottles, tin cans, plastic bags – precisely which objects serve as play objects is ultimately irrelevant. Thus, toys produced through artisan or industrial processes are simply one special form of toys that, as mentioned above, are closely tied to the development of bourgeois notions about what being a child and what family means. Initially Germany was the centre of the toy industry, with the heart of production shifting later to the USA, until, beginning from the second half of the 20th century, China became the market leader.

Although children's play, seen from the outside, might appear to be a rather purposeless drift, this drift actually fulfils a deeper purpose that is ultimately necessary for survival: It enables a form of activity that is guided by curiosity, a willingness to learn and imagination. Play is motivated and rewarded by the sensation of fun and the feeling of 'flow' that the child experiences – a feeling of being able to lose oneself in the activity at hand (Csíkszentmihályi 2000).

ludens, den Menschen, der im Spiel zum Wesen seiner selbst findet. Freiheit jenseits von Notwendigkeit und Pflicht, Leichtigkeit des Geistes, Schönheit, Genuss von Kreativität – das sind Akzente, die Schiller setzte und die seither in der philosophischen und kulturwissenschaftlichen Spielforschung diskutiert werden.

Mit Friedrich Schiller eine Flüchtlingsunterkunft zu besuchen und mit seinen Augen Kinder beim Spielen zu beobachten, ist eine Übung der besonderen Art. Spielen mit Flucht und Migration zusammenzubringen, fällt schwer. Dreck, Nässe, Kälte, Gewalt, Ausgeliefertsein und ungewisses Warten – all dies schafft das genaue Gegenteil einer Atmosphäre von Freiheit und Leichtigkeit des Geistes, die doch Voraussetzung für das spielerische Handeln sein soll. Nichtsdestotrotz sind auch dort überall spielende Kinder anzutreffen. Ist es also nicht umgekehrt – Freiheit und innere Leichtigkeit nicht als Voraussetzung von Spiel, sondern als dessen Ergebnis? Kann die Magie des Spiels Stacheldraht und Betonmauern vergessen lassen?

Was für Erwachsene eine Wirklichkeit voller Entbehrung, Erniedrigung, Schrecken und Angst darstellt, verwandelt sich im kindlichen Spiel in ganz andere Szenarien. Das völlig zerbombte Nürnberg des Jahres 1945 war für Kinder »der schönste Spielplatz der Welt«, so berichten Zeitzeugen (Falkenberg 2015: 10). Die ersten Luftangriffe auf Berlin wurden in kindlicher Wahrnehmung zu einer Sensation und Bombensplitter zu begehrten Tauschobjekten, wie sich der spätere Drehbuchautor Wolfgang Kohlhaase (* 1931) erinnert (Dieckmann 2021, siehe

The instrumental perspective applied in behavioural research and educational science stands in contrast with a philosophical valuing of play. The latter does not emphasise the functions that play unquestionably contains regarding psychological development, but rather the utopian potential of play. "The human being only plays in places where they are human in the full sense of that word, and they are only entirely human in places where they play." (Schiller 1795: 88; translation Henry Holland).

In this now famous aphorism, Friedrich Schiller celebrated the *Homo ludens*, the playing human, who finds their way to what is essential in themselves through play. Freedom beyond any forms of necessity and duty, an easiness of the intellect, beauty, pleasure in creativity – these are the elements Schiller accentuates in his account of play, and which have been discussed in research and discourse on play to this day, as conducted by philosophers and cultural studies scholars.

Visiting refugee accommodations as if Schiller were accompanying you on your visit and observing children at play with his eyes is a very particular type of creative and intellectual exercise. People find it hard to connect playing with flight and migration. Dirt, damp, cold, violence, being thrown at the mercy of fortune and uncertainty while waiting: All these factors establish an atmosphere that is the exact opposite of freedom and ease of the spirit, which, however, is meant to be the precondition for playful conduct – at least according to Schiller. This notwithstanding, those who undertake such a visit come across chil-

Abb. 2: Flüchtlingskinder spielen mit einem Stofftier in einem behelfsmäßigen Lager an der griechisch-mazedonischen Grenze in der Nähe des Dorfes Idomeni. Die Aufnahme wurde am 15. März 2016 gemacht. Wenige Tage zuvor hatte Mazedonien seine Grenze geschlossen. Alle Migrant:innen wurden nach Griechenland zurückgeschickt. Die Balkanroute sollte zu diesem Zeitpunkt unter allen Umständen blockiert bleiben, wie der damalige österreichische Verteidigungsminister Hans Peter Doskozil verlautbaren ließ. In den behelfsmäßigen Auffanglagern befanden sich zu diesem Zeitpunkt 12.000 Flüchtende. Wie das unhandliche und wenig wetterfeste Stofftier in das Lager gekommen ist, wird nicht berichtet. Ebenso ungewiss bleibt sein weiteres Schicksal und das der mit ihm spielenden Kinder. ©Alexandros Avramidis/Reuters

Image 2: Refugee children playing with a teddy bear in a makeshift camp on the Greek-Macedonian border near the village of Idomeni. The photo was taken on 15 March 2016. Macedonia had closed its border only days previously. All migrants were sent back to Greece. European decision-makers at the time wanted to close down the Balkan Route at all costs, a goal reflected in statements made by the then Austrian Defence Minister, Hans Peter Doskozil. Around 12,000 refugees were housed in the makeshift reception camps when the border was closed. There are no reports as to how the unwieldy and not very weather-resistant teddy got into the camp. Its future fortune is equally uncertain, as is the future of the children playing with the teddy. ©Alexandros Avramidis/Reuters

auch Landgraf und Pfirschke 2005).[2] In den zahllosen Flüchtlingslagern rund um den Globus leben Kinder unter häufig entwürdigenden Zuständen. Trotzdem spielen sie und erleben Spaß. Diese Behauptung weckt widerstrebende Reflexe. Es liegt mir nicht daran, die schillersche Idealisierung des Spiels fortzuschreiben. Ohne Frage erleben Kinder auf der Flucht und in Lagern Leid, auch schwer erträgliche Langeweile und dren playing everywhere. Is it, perhaps, not the other way around – that freedom and inner ease are not the precondition of play, but rather its result? Can the magic of play make it possible for children to forget barbed-wire and concrete walls?

An environment that represents, for adults, a reality full of want, humiliation, panic and anxiety transforms itself in children's play into entirely different scenes.

Toys

depressive Stimmungen. Es gibt Berichte, wonach im Flüchtlingslager Moria in Griechenland Kinder versuchten, sich das Leben zu nehmen. Die desaströse äußere Situation führte dort zu einer ebenso desaströsen inneren Verfasstheit und zeigt auf, was passiert, wenn Kindsein nicht im Spiel gelebt werden kann.[3] Dort, wo Spiel unmöglich ist, wird es ernst. Gleichzeitig und gerade deswegen gilt es, die ganz eigene kindliche Weltwahrnehmung im Spiel anzuerkennen und ihr wirklichkeitsstiftendes Potenzial ernst zu nehmen.

Puppen

Die Puppe begleitet Menschen nachweislich seit der Jungsteinzeit. Aus Ton gefertigte Figürchen in Kindergräbern im östlichen Mittelmeerraum belegen dies. Archäolog:innen diskutieren ihre Vieldeutigkeit. Handelte es sich ›nur‹ um Spielzeug oder Darstellungen von (Schutz-)Göttinnen oder um Figurinen, die im Rahmen von Initiations-, Heilungs- oder Fruchtbarkeitsriten gebraucht wurden (Talalay 1993)? Festzuhalten ist: Für das Mensch-Puppen-Verhältnis ist Vieldeutigkeit ein grundsätzlicher Wesenszug. Das gilt insbesondere für Puppen, die massenhaft ab dem 19. Jahrhundert zunächst in Franken und Thüringen gefertigt wurden und mittlerweile in Ostasien produziert werden. Allein die variantenreiche Typologie spricht für unterschiedliche Verwendungsweisen: Künstlerpuppe, Modepuppe, Gliederpuppe, russische Matrjoschka, Spielpuppe, Babypuppe, sprechende Puppe, trinkende Puppe, Barbie-Puppe …

Industriell gefertigte Spielzeugpuppen

Eyewitnesses report that the utterly bombed-out city of Nuremberg in 1945 was "the loveliest playground in the world" for children (Falkenberg 2015: 10). In children's observations, the first air attacks on Berlin were transformed into something sensational, and bomb fragments became objects of desire for swapping – as the screenwriter Wolfgang Kohlhaase, born 1931, later remembered (Dieckmann 2021; see also Landgraf and Pfirschke 2005).[2] In countless refugee camps across the globe, children often live under degrading circumstances. But, despite everything, they manage to play and have fun. While such an assertion may be greeted antagonistically, in a reflex manner, my aim here is not to write a sequel to Schiller's idealisation of play. It's undisputed that children who are fleeing or living in camps experience suffering, including almost unbearable boredom and depressive moods. There are reports that children have attempted to commit suicide in the Moria Refugee Camp in Greece. The disastrous external situation in Moria leads to an equally disastrous inner state and highlights what can happen when being a child cannot be lived through play.[3] In places where play is impossible, the situation becomes extremely serious. Concurrently, and for the very same reason, the imperative is to recognise children's entirely particular perception of the world through play and to take seriously the potential this has to (re)generate reality.

sind überwiegend weiblichen Geschlechts, meistens im Baby- oder Kleinkindalter.[4] Sie tragen mehrheitlich weiße Hautfarbe und mitteleuropäische Gesichtszüge. Ausnahmen stellen schwarze Babypuppen dar, deren früheste Exemplare sich in den Südstaaten der USA im 18. Jahrhundert finden lassen und die seit dem späten 19. Jahrhundert im Sortiment aller renommierten Puppenhersteller enthalten sind. Puppen mit asiatischen oder auch lateinamerikanischen Gesichtszügen sind erst seit kurzem auf dem Markt und stellen (noch) nicht den Standard dar. Wenn flüchtende Kinder aus Afrika, Lateinamerika, dem Nahen oder Mittleren Osten mit Puppen angetroffen werden, tragen diese Gesichter von weißen Kindern.

Dolls

It is known that dolls have accompanied human life since the Neolithic period. Small clay figures in children's graves in the Eastern Mediterranean area are testimony to this. And archaeologists are engaged in an ongoing discussion on what they may have signified and what their function may have been. Were they 'merely' toys or representations of protective female Gods, needed within the context of initiation, healing or fertility rituals? (Talalay 1993) What we do know is that multiplicity is a fundamental part of the essence of human-doll relations. This is especially true for dolls that were mass produced from

Abb. 3: Ein kongolesisches Mädchen, das 2013 durch die Kämpfe in Nord-Kivu vertrieben wurde und sich in einem Lager bei Goma (Demokratische Republik Kongo) befindet, lässt sich mit ihrer Puppe fotografieren. Ob sie zusammen mit ihrer Puppe vertrieben wurde oder ob die beiden sich erst im Lager kennengelernt haben? ©Thomas Mukoya/Reuters

Image 3: A Congolese girl, driven from her home by the 2013 battles in the Nord Kivu region of the country and, at the time this image was created, living in a camp near Goma (also in Congo), has her photo taken alongside her doll. Whether she was forced to flee with her doll or whether they both only got to know each other in the camp is unknown. ©Thomas Mukoya/Reuters

Abb. 4: Das kurdische Flüchtlingsmädchen aus der syrischen Stadt Kobanê sitzt vor seinem Zelt in einem Flüchtlingslager bei Suruç (Provinz Şanlıurfa). 24. November 2014. ©Osman Orsal/Reuters

Image 4: This Kurdish refugee girl from the Kurdish-majority city of Kobanî in Syria sits in front of its tent in a refugee camp near Suruç, in the province of Şanlıurfa. 24 November 2014. ©Osman Orsal/Reuters

Die Puppe ist eine Person, mitunter sogar eine Persönlichkeit, die einen Namen trägt, mit der gesprochen wird, die auch Widerworte geben oder traurig sein kann und getröstet werden muss. Das Puppenwesen gibt zu erkennen, dass es bedürftig ist, es will umsorgt sein. Dieses Artefakt ist auf eigene Weise ›beseelt‹, und es spricht vor allem Mädchen an. Gleich ob dahinter gendernormative Erziehung vermutet wird oder biologische Anlagen, Studien zur Spielzeugpräferenz, die größtenteils in den USA, Kanada, Großbritannien und Australien durchgeführt wurden, deuten darauf hin, dass vor allem Mädchen mit Puppen spielen, vermuten aber auch, dass es hier kulturelle Unterschiede gibt (Davis und Hines 2020: 387). Im Umgang mit der

the 19th century on, initially in Franconia and Thuringia, and now primarily in the Far East. The multifaceted typology of dolls is itself a record of their very varied utilization: artists' dummies, fashion dolls, mannequins, Russian Matryoshka dolls, dolls for playing, baby dolls, speaking dolls, drinking dolls, Barbie dolls …

Industrially produced toy dolls are overwhelmingly made to look female and are mostly meant to represent babies or toddlers.[4] They are mostly made with white skin and Central European facial features. Black baby dolls are a longstanding exception, the first examples of which were to be found in the USA's southern states in the 18th century and which have been

Abb. 5: Die Mutter trägt ihre erschöpfte Tochter, die Schwester trägt ihre Puppe, die mit einem Schnuller beruhigt werden muss. Sie und andere Migrant:innen sind nach ihrem Grenzübertritt aus Mazedonien auf dem Weg nach Miratovac, Serbien, im Oktober 2015. ©Marko Djurica/Reuters

Image 5: The mother is carrying her exhausted daughter, while her sister carries her doll, which is being calmed down with a pacifier. Together with other migrants, they are making their way to Miratovac, Serbia, in October 2015, after crossing the border out of Macedonia. ©Marko Djurica/Reuters

Puppe lassen sich unschwer elterliche Verhaltensweisen wiederentdecken: Babys wiegen, Trost schenken, beruhigen, baden, wickeln und herzen. Das Kind-Puppen-Spiel spiegelt Schutzbedürftigkeit einerseits und andererseits das Gefühl, gebraucht zu werden, welches seinerseits Stärke verleiht.

Kinder, die auf Fluchtrouten unterwegs und in Lagern festgesetzt sind, sind in hohem Maße schutz- und trostbedürftig. Vielleicht verleiht das Spiel mit der schutzbedürftigen Puppe Kraft? Vielleicht ist hier der Begriff ›Spiel‹ aber auch der völlig falsche?

part of the product range of renowned doll manufacturers since the late 19th century. Dolls with Asian or South American facial features have only come onto the global market recently and are not yet in any way the standard. When children fleeing out of Africa, South America or the Near or Middle East are welcomed with dolls, it's highly likely that these dolls have the faces of white children.

The doll is a person, sometimes even a personality, bearing a name, who can be spoken to, who can also talk back and who can be sad and needs to be comforted. The doll as a being makes it known that it is needy and that it wants to be cared for. This artefact is, in its own unique way 'ensouled'

Bären

Der Teddybär ist weltweit bekannt und geliebt. Im Gegensatz zur Puppe ist er allerdings noch nicht sehr alt, gut einhundert Jahre sind seit seiner Geburt vergangen. Die Geburtsumstände sind in zwei Versionen überliefert (Koenneritz 1947; Cockrill 1994). In der deutschen Version wurde der erste Plüschbär mit beweglichen Armen und Beinen 1902 von dem Erfinder Richard Steiff (1877–1939), Neffe der bekannten Spielzeugherstellerin Margarete Steiff, erschaffen und erhielt den wenig klangvollen Namen Bär 55 PB (P steht für Plüsch, B für beweglich). Der 55 cm hohe Plüschbär, der in Giengen an der Brenz (Baden-Württemberg) das Licht der Welt erblickte, wurde 1903 auf der Spielwarenmesse in Leipzig dem Fachpublikum präsentiert, stieß dort jedoch auf wenig Interesse. Kurz vor Schließung der Messe, so heißt es, kaufte ein Vertreter des amerikanischen Spielzeugwarenvertriebs FAO Schwarz jedoch einhundert Exemplare. Die Legende will es, dass ein solcher Steiff-Bär vom Sekretär des amerikanischen Präsidenten Theodore »Teddy« Roosevelt (1859–1919) als Verlegenheitsgeschenk für den Geburtstag von dessen Tochter erworben wurde. Das begeisterte Kind taufte das Tier nach seinem Vater »Teddy«. Bereits 1904, auf der Weltausstellung in St. Louis, wurden 12.000 Bären verkauft und Margarete und Richard Steiff mit der Goldmedaille der Weltausstellung ausgezeichnet.

Die amerikanische Version verbindet die Entstehung des Teddybären mit einem Jagdausflug des Präsidenten Roosevelt. Als passionierter Jäger wollte er im November 1902 in Mississippi einen Bären erlegen. Ein

and appeals principally to girls. Irrespective of whether this is attributed to gender-normative upbringing or to biological predisposition, studies on toy preference, conducted primarily in the USA, Canada, the UK and Australia, point out that it is girls who play most with dolls, but also surmise that there are cultural differences in this regard (Davis and Hines 2020: 387). How children relate to dolls makes it easy to recognise parental behaviour patterns: rocking babies, comforting, calming, changing nappies and clothes, and hugging. Play between a child and a doll reflects a need to be protected, on the one hand, and a feeling of being needed, on the other hand – an emotion that strengthens the child psychologically.

Children who are fleeing on refugee routes or who are stuck in camps greatly require protection and comfort. Perhaps playing with a doll in need of protection can give this child strength. Alternatively, is it also possible that the very concept of 'play' is wrong entirely in this context?

Bears

The teddy bear is known and beloved all over the world. In contrast to the doll, however, the bear is not very old, with slightly over a century now passed since its birth. Two different stories about the conditions of its birth have been handed down to us (Koenneritz 1947; Cockrill 1994). In the German version, the first furry toy bear with moving arms and legs was created in 1902 by the inventor Richard Steiff (1877–1939), a nephew of the renowned toy-maker

solcher war jedoch nicht gesichtet worden. Eingefangen wurde lediglich ein kleines mutterloses Tier. Von einem Jagdhelfer an einen Baum gebunden, sollte Roosevelt es erschießen, was dieser verweigerte. Ein Karikaturist machte aus der Begebenheit ein Titelbild für die Ausgabe der *Washington Post* vom 16. November 1902. Die Episode wurde im Folgenden wiederholt ins Bild gesetzt, wobei der Bär immer runder und niedlicher geriet und alsbald als »Teddys Bär« bekannt wurde und symbolisch für den Präsidenten stand.

Davon angeregt bastelte der russisch-jüdische Immigrant Morris Michtom (1870–1938) mit Hilfe seiner Frau Rose einen Bären und stellte ihn als Dekoration ins Schaufenster seines Ladens *Ideal Novelty and Toy Company* in Brooklyn. Roosevelt soll ihm schriftlich die Erlaubnis erteilt haben, das Tier als »Teddy's Bear« zu bezeichnen. Das Ehepaar Michtom verkaufte alsbald Bären mit beweglichen Gliedmaßen und gleichzeitig das positive Image des Tierfreundes Theodore Roosevelt.

Es dürfte müßig sein, sich entweder für Morris Michtom oder Richard Steiff als Urheber auszusprechen. Der Siegeszug des Plüschtiers verdankt sich wohl beiden. Die Popularität des bis dahin jüngsten Präsidenten, der sich weigerte, einen putzigen kleinen Bären zu erschießen, und das gut situierte Bürgertum der Ostküste der Vereinigten Staaten, dessen Erziehungsstil eine erhöhte Nachfrage an Spielzeug weckte, beförderten diese Erfolgsgeschichte maßgeblich.

Die sensationelle Karriere des Teddys ist auch als Migrationsgeschichte zu verstehen. Als Bär 55 PB reiste er 1903 von Giengen nach Baltimore und wurde dort in der Neuen Welt als Migrant bekannt und schließlich weltweit

Margarete Steiff, and was given the not very melodious name of *Bär* [Bear] 55 PB (P stood for *plüsch*, meaning furry, and B für *beweglich*, moveable). This 55 cm-high furry bear, who entered this world in Giengen-an-der-Brenz, in what is now the state of Baden-Württemberg), was presented to experts in the industry at the toy-fair in Leipzig in 1903, but generated little interest. However, shortly before the fair was about to close, so the story goes, a rep from the American toy brand FAO Schwarz bought up 100 of these toy bears. As legend will have it, one of these very Steiff bears was then purchased by the secretary of the US President Theodore "Teddy" Roosevelt (1859–1919) as a last-minute birthday present for the president's daughter. The delighted child christened the animal "Teddy" after her father. At the St. Louis World's Fair in 1904, 12,000 bears were sold already, and Margarete and Richard Steiff were awarded the fair's gold medal.

The American version of the origins of the teddy bear connects the emergence of that figure with one of President Roosevelt's hunting trips. A passionate hunter, Roosevelt wanted to shoot a bear in November 1902 in Mississippi. But no such bear could be sighted. All the president's gamekeeping assistants were able to capture was a small, motherless infant bear, which was tied to a tree by one of the assistants so that Roosevelt could shoot it. But the president refused to do so, and a caricaturist's drawing of the scene made the front page of the *Washington Post* on 16 November 1902. Many other illustrations of the episode followed, with the bear

Toys

Abb. 6: »Drawing the Line in Mississippi«. Der leidenschaftliche Jäger Theodore Roosevelt weigert sich, den Bären zu erschießen. Zwar wurde dieser dennoch, von Roosevelts Begleiter John M. Parker, mit dem Messer getötet, erlangte jedoch als Teddybär eine besondere Form der Unsterblichkeit. Karikatur von Clifford K. Berryman in der Washington Post, 16. November 1902 (gemeinfreie Abb.)

Image 6: "Drawing the Line in Mississippi." The passionate hunter Theodore Roosevelt refuses to shoot the bear. Although subsequently killed with a knife by Roosevelt's companion, John M. Parker, the teddy bear acquired its own form of immortality nonetheless. Caricature by Clifford K. Berryman in the Washington Post, 16 November 1902 (creative commons license).

geschätzt. Über den Spielwarengroßhandel des deutschen Migranten Frederick August Otto Schwarz (1836–1911) und über den Laden der russischen Immigranten Michtom gelangt er in den Verkauf und zur Tochter des amtierenden Präsidenten. Seit 1906 ist »Teddybär« die handelsübliche Bezeichnung des Stofftiers und bis heute ist der 9. September in den USA der *National Teddy Bear Day*.

Im Gegensatz zur Puppe ist der Umgang mit dem Teddy an kein Geschlecht gebunden und wird nicht selten bis ins Erwachsenenalter gepflegt. Er ist, mehr noch als die Puppe, biografisches Objekt und Lebensbegleiter. Seine Artenvielfalt ist im Laufe seiner hundertjährigen Geschichte immens gewachsen und beflügelt Sammelleidenschaften.

seemingly becoming rounder and sweeter in each one, before it soon became known as "Teddy's bear" – and was seen as standing symbolically for the president himself.

Inspired by this turn of events, the Russian-Jewish immigrant Morris Michtom (1870–1938) and his wife Rose made their own bear and placed it as a decoration in the window of their store, *Ideal Novelty and Toy Company*, in Brooklyn. Roosevelt then gave him written permission to call the toy animal "Teddy's Bear". Soon the Michtoms were selling such bears with moveable limbs – and were simultaneously promoting a positive image of Theodore Roosevelt as an animal lover.

It would be pointless to pronounce that either Morris Michtom or Richard Steiff were the exclusive inventors of this novelty. Both should be thanked for the furry animal's march to victory. The popularity of a man who was, at that time, the youngest president to date and who refused to shoot a small, sweet bear, combined with a well-off middle-class in the East coast of the USA, whose child-rearing methods led to an increased demand for toys, to shape this success story substantially.

The teddy bear's sensational career should also be understood as a story of migration. Still known at that point in time as *Bär 55 PB*, he journeyed from Giegen to Baltimore in 1903, found fame in the New World as a migrant and is ultimately valued everywhere. Via the toy wholesale business of the German immigrant Frederick August Otto Schwarz (1836–1911), and via the toy store run by the Michtoms, Russian immigrants, he came into economic

In Zeiten von Verunsicherung und turbulenten äußeren Bewegungen wird der handliche Bär mit den treuen Glasaugen zum personifizierten ruhenden Pol. Seine zuverlässige körperliche Nähe gibt kleinen (und größeren) Menschen die Möglichkeit, Übergangsphasen zu bewältigen und Weltvertrauen zu erwerben. Wärme, Geborgenheit, Liebe, Stabilität – all diese elementaren seelischen Grundbedürfnisse kann das mit Polyesterwatte (früher Holzwolle) gefüllte Artefakt aus Plüsch und Filz erfüllen. Teddybären können für Kinder gar lebenswichtig sein, wenn sie helfen, Ängste oder Trennungsschmerz zu lindern (Bosch 2014).

circulation and from there into the hands of the daughter of a sitting president. "Teddy bear" has been the normal trade description of this cuddly toy since 1906, and September 9 remains *National Teddy Bear Day* in the USA until this day.

In contrast to the doll, no clear gender patterns are evident in interactions with teddies, and relationships with them are often nurtured into adulthood. A teddy is, more than a doll, a biographical object and a lifelong-companion. The diversity of the teddy as a 'species' has grown immensely during its 100-year history and has lent wings to collectors' passions.

Abb. 7: Ein geflüchtetes afghanisches Mädchen überreicht der Freundin ihren Teddybären. Beide teilen dessen Zuneigung. Islamabad, 14. Februar 2014. ©Sara Farid/Reuters

Image 7: An Afghan refugee girl hands over her teddy bear to her friend. Both share in its affections. Islamabad, 14 February 2014. ©Sara Farid/Reuters

Es verwundert nicht, dass in Zeiten kollektiver Krisen wie Krieg, Vertreibung und Flucht auch der Teddybär als zuverlässiger Begleiter an der Seite von Kindern zu finden ist (vgl. z.B. Falkenberg 2015: 168–175; Landgraf und Pfirschke 2005; Seidel 2006).

Drachen

Ein Tier völlig anderer Art stellt der Drache dar. Als fliegendes Fabelwesen, einem Reptil ähnelnd, ist es in östlichen und westlichen Mythen, Märchen und Legenden überliefert und zudem als Sternbild und Wappentier bekannt (Canby 1997; Schmelz und Vossen 1995). Flugobjekte mit einem mythisch-reli-

In times of uncertainty and turbulent external movements, the convenient-sized bear with its loyal glass eyes becomes the personification of serenity. Its reliable bodily proximity gives both young and old a means with which they can overcome transitional phases in their lives, and acquire trust in the world. The foundational, psychological needs of warmth, emotional security, love and stability can be fulfilled by this artefact, made from fur or felt-type materials, and stuffed today with polyester filling – formerly filled with wood shavings. Teddy bears can even be essential in some children's' lives, when they help to mitigate anxieties or the pains of separation (Bosch 2014).

It is no surprise that teddies are to be found, as reliable companions, beside children, in times of collective crises, including war, persecution and forced flight (cf. Falkenberg 2015: 168–175; Landgraf and Pfirschke 2005; Seidel 2006).

Kites and Dragons

A dragon is an entirely different sort of beast. Stories about dragons as flying-beings from fables, similar to a reptile, have been passed down to us through Eastern and Western myths, fairy tales and legends are also known to us through the constellation Draco, the Latin word for dragon, and as a creature adorning coats-of-arms (Canby 1997; Schmelz and Vossen 1995).[5] Flying objects with mythical and religious connections to both dragons and ritual usage came into being in East and Southeast

Abb. 8: Flüchtlingskinder der Rohingya lassen vom Dach einer Behelfsunterkunft Drachen steigen. Die Kinder leben im Lager Balikhali in Cox's Bazar (Bangladesch). 14. November 2018. ©Mohammad Ponir Hossain/Reuters

Image 8: Refugee children from the Rohingya people fly a kite from the roof of a makeshift shelter. The children live in the Balikhali Camp in Cox's Bazar, Bangladesh. 14 November 2018. ©Mohammad Ponir Hossain/Reuters

Abb. 9: Spaß an selbst gebauten Drachen. Die Geflüchteten befinden sich im sogenannten ›Dschungel von Calais‹, einer provisorischen Zeltstadt in der Nähe von Calais (Frankreich). Die Aufnahme wurde am 12. Oktober 2016 gemacht. Seit Anfang 2016 fanden regelmäßige Räumungsversuche durch die Polizei statt. Mittlerweile ist das Lager komplett zerstört. ©Chebil

Image 9: Fun with homemade kites/'dragons'. The refugees in the photo used to live in the Calais Jungle, the ramshackle tent-city near Calais, France. The photo was taken on 12 October 2016. Regular attempts to evict the refugees were carried out by the police from early 2016. The camp has now been completely destroyed. ©Chebil

giösen Drachenbezug und ritueller Verwendung sind in Ost- und Südostasien vor mehr als 2500 Jahren entstanden. Über portugiesische, holländische und englische Händler kamen Papierdrachen ab dem 16. Jahrhundert nach Europa und fanden ab dem frühen 18. Jahrhundert als Spielzeug Verwendung. Die im Deutschen geläufige Bezeichnung ›Drache‹ für solch ein Flugobjekt stammt vom römischen Militär. Ein Feldzeichen, die Draco-Standarte, stellte einen stilisierten Drachenkopf dar, an dem eine im Wind flatternde Tuchröhre befestigt war.

Drachen steigen zu lassen, ist auch mit einfachsten Mitteln möglich. Benötigt werden ein Gestänge aus Holz, Bambus oder

Asia more than 2500 years ago. Portuguese, Dutch and English merchants brought such paper kites to Europe from the 16th century, objects that were used as toys from the early 18th century. The German use of the word *Drache* [dragon] for this sort of flying object can be traced back to the Roman military. The so-called *Draco* standard, a battle symbol, consisted of a stylised dragon's head, onto which was fixed a roll of cloth that flattered in the wind.

It is possible to fly kites using the simplest of means. Flyers require a frame made of wood, bamboo or metal, a sail that doesn't let air through, mostly made from synthetic fabric, cotton, silk or paper, and

Toys

Metall, ein luftundurchlässiges Segel, meist aus Kunststoff, Baumwolle, Seide oder Papier, und eine lange Leine, die die Verbindung zum Piloten herstellt. Neben dem Material und ein wenig Bastelgeschick ist noch etwas anderes unabdingbar: der Wind. Das Spiel mit einem Drachen ist das Spiel mit dem luftigen Element. Während Puppe und Bär eng mit dem eigenen Körper, mit Intimität, Berührung und vertrauensvoller Kommunikation verbunden sind, funktioniert das Spiel mit dem Drachen nur auf Distanz. Das Fluggerät muss an der langen Leine belassen werden, und je höher er aufsteigt, desto größer der Spaß. Eingeübt wird auch das kunstfertige Loslassen-Können. Die Assoziation mit flugfähigen Tieren drängt sich auf: Mit dem Drachen schwebt ein Teil des kindlichen Selbst ›frei wie ein Vogel‹ weit über den einengenden Beschwernissen der irdischen Existenz.

a long string, which facilitates the connection between the kite and the kite flyer. Besides these materials and a little dexterity in making things, another element is indispensable: the wind. Playing with the kite or 'dragon' means playing with the airy element.

While dolls and teddies are closely connected with their bearers' own bodies, with intimacy, touch and trustworthy communication, playing with the kite/dragon only works at a distance. The flying machine must be let out on a long cord, and the higher it climbs, the greater the fun. This is also practice in the art of being able to let go. Associations with animals that can fly are emphasised: along with the kite/dragon, a part of the child's self floats 'free as a bird' far above the restrictive burdens of earthly existence.

Waffen

Weapons

Militärisches Blech-, Blei- und Zinnspielzeug fanden sich in den Warenkatalogen der Spielwarenhersteller bereits im frühen 19. Jahrhundert. Die Herausbildung europäischer Nationalstaaten war verbunden mit der Bewaffnung der männlichen Bevölkerung im verpflichtenden Wehrdienst. Das Soldatische wurde in Folge der Militarisierung europäischer Nationalstaaten zum Männerideal. Die bürgerliche Erziehung orientierte sich selbstredend an solchen normierten Geschlechterrollen. Dahinter stand die Idee, dass Mädchen von alleine, also ›natürlich‹, zur Frau reifen, Jungs aber zu ›echten‹ Männern geformt werden müssen. Auf Europa bezogen und historisch

Military toys made from sheet metal, tin or lead were to be found in the catalogues of the toy manufacturers from the early 19th century. The formation of European nation states went hand in hand with the arming of the male population in the course of compulsory military service. One consequence of the militarisation of European nation states was that the soldier became the ideal of manhood. It appeared self-evident that bourgeois education should orient itself towards such normed gender roles. Such an education was backed by the idea that girls could mature into women by themselves, or 'naturally', whereas boys

betrachtet, fand seit dem 19. Jahrhundert, vor allem aufgrund der kriegerischen Erfolge Preußens, eine Militarisierung der Kinderstube statt, die nun fast unvermeidlich mit Zinnsoldaten, Schaukelpferd, Trommel, Säbel und Gewehr bestückt wurde. Zackige Bewegungen, Salutieren und Exerzieren schrieben sich in den Körper des männlichen Kindes ein. Der Krieg von 1871, der Erste Weltkrieg, Nationalsozialismus und Zweiter Weltkrieg beförderten das Kriegsspiel nachhaltig (Kroner 1982; Greim 2004: 17f.). Dahinter steht die These, dass seit über einhundert Jahren die westliche Moderne ihre gewalthaltige Seite in die Kinderzimmer außereuropäischer Regionen exportierte. Entwicklungspsychologisch betrachtet, kann aber auch behauptet werden, dass das Zusammenspiel von Mann und Waffe eine anthropologische Konstante darstellt, also universell und die Waffenleidenschaft somit biologisch verankert sei. Diese Fragen sind hier nicht zu beantworten.

Fest steht: Spielzeugwaffen in unterschiedlich realistischen Varianten finden sich bis heute in vielen Kinderzimmern, weltweit. Kampagnen gegen Kriegsspielzeug bzw. ›Gewaltspielzeug‹ sind nicht ungewöhnlich und nicht nur auf die westliche Mittelschicht beschränkt. Kleine Jungs in Tarnanzügen, die sich mit Handgranaten, Maschinenpistolen, Sturmgewehren, Pumpguns, Revolver, Panzern und Bombenflugzeugen die Zeit vertreiben, gelten aus dieser Perspektive als potenziell gefährdet und gefährlich. Andererseits produziert die Spielzeugindustrie, im Verbund mit Hollywood-Blockbustern, hypermaskuline Heldenfiguren, die muskelbepackt und mit Hightech-Waffen ausgestattet den Nimbus der Unbesiegbarkeit in sich tragen.

had to be formed into 'real' men. With a focus on Europe and regarded with historical hindsight, we can say that a militarisation of the nursery took place from the 19th century, with Prussia's successes in war centre-stage in this process. In this era, it seemed almost impossible to avoid equipping the nursery with tin soldiers, rocking horses, drums, sabres and rifles. Snappy movements, salutes and military drill wrote their way into the bodies of male children. The Franco-Prussian War of 1870–1871, the First World War, National Socialism and the Second World War all lent sustained impetus to the game of war (Kroner 1982; Greim 2004: 17f.). These conclusions are supported by the thesis that Western modernity has been exporting its violent side to children's bedrooms outside Europe for over 100 years. In the language of developmental psychology, it can be claimed that the interaction between men and weapons represents an anthropological constant: The argument is that this interaction is universal, and that a passion for weapons arises from a biological predisposition. No answer is given to these claims here.

But it is indisputable that toy weapons, inconsistently realistic in their appearance, are a presence in many children's playrooms and bedrooms around the world. Campaigns against war toys, or 'violent toys' as they're sometimes known, are no rarity, and their reach extends beyond the Western middle class. From the campaigners' perspective, little boys wearing camouflage and killing time with the help of grenades, submachine guns, assault rifles, pump-action shotguns, revolvers,

Toys

Abb. 10: Zwei Jungen in einem Flüchtlingslager in Bagdad nach dem zweiten Irakkrieg. Die Waffen sind nicht echt, sondern nur Spielzeugpistolen. Vorausgegangen war die Flucht von Tausenden von Einwohnern aus der umkämpften Stadt Faludscha, nachdem die US-Militärs eine Offensive angekündigt hatten (November 2004). ©Namir Noor-Eldeen/Reuters

Image 10: Two boys in a refugee camp in Baghdad after the 2003 invasion of Iraq. The weapons are toy pistols, not real guns. Prior to this photo, thousands of residents had fled from the embattled city of Fallujah, after the US military had announced an offensive (November 2004). ©Namir Noor-Eldeen/Reuters

Diese Actionfiguren erobern weltweit Kinderherzen. Action und Spielspaß kreisen dabei um ›Kämpfen‹ und ›Zerstören‹, was weitgehend ein männliches Phänomen darstellt, wohingegen »Mädchen eher niedliche Tierfiguren aus Cartoons oder Animes als Helden einsetzen«, wie Ingrid Paus-Hasebrink feststellt (2010: 4) Das von Kindern bevorzugte Spielzeug spiegelt auch kulturell verankerte Geschlechterrollen.

Vielleicht haben Superman und Batman,

tanks and bomber planes are viewed as being potentially in danger *and* potentially dangerous. Juxtapose this with the toy industry producers, who, in conjunction with Hollywood business leaders, pump out hypermasculine hero figures. With their rippling muscles and equipped with high-tech weapons, these industrial creations seem to be surrounded by a glow of invincibility. These action figures conquer children's hearts globally. Action and joy in play using such toys circle around the activities of 'fighting' and 'destroying', activities that enact a male phenomenon to a large degree. Whereas, according to Ingrid Paus-Hasebrink's research (2010: 4), "girls [are more likely to] make heroes out of cute animal characters from cartoons or animes" in their play, the toys children prefer reflect culturally-rooted gender roles.

Perhaps Superman, Batman and Wonder Woman, those saviours of the world in *The Avengers*, or the protagonists in games like *Fortnite* have long since replaced the teddy bear as companions who can provide protection and comfort. The phantasm of having one's own hidden powers and the identification with invincible heroes and heroines are often part of a certain developmental phase in which the experience of powerlessness demands compensation. Miniature plastic heroes and heroines as well as larger-than-life actors on the movie screen nourish precisely such needs and fantasies.

Living through war, flight and expulsions, children are confronted with real bombs, tanks and submachine guns, and with adults who carry real weapons in their

Abb. 11: Vertriebene Kinder im Lager Deepaka nordwestlich von Erbil (Irak). Die Jungen sind vor den Kämpfern des Islamischen Staates aus Mossul geflohen und spielen das Erschießen von Feinden. 20. Oktober 2016. ©Alaa Al-Marjani/Reuters

Image 11: Forcibly expelled children in the Deepaka Camp northwest of Erbil, Iraq. Islamic State fighters caused the boys to flee from Mosul, and they now play shooting enemies. 20 October 2016. ©Alaa Al-Marjani/Reuters

Wonderwoman, die Weltretter:innen der *Avengers*-Serie oder auch die Protagonist:innen von Spielen wie *Fortnite* längst den Teddybären als Schutz- und Trostgefährten ersetzt? Das Phantasma verborgener eigener Macht und die Identifikation mit unbezwingbaren Held:innen sind häufig Teil einer bestimmten Entwicklungsphase, in der die Erfahrung eigener Ohnmacht nach Kompensation verlangt. Kleinformatige Plastikheld:innen ebenso wie großformatige Akteur:innen auf der Filmleinwand nähren genau solche Bedürfnisse und Fantasiewelten.

Krieg, Flucht und Vertreibung konfrontieren Kinder mit echten Bomben, Panzern, Maschinenpistolen und mit Erwachsenen, die echte Waffen in ihren Händen tragen und diese auch gebrauchen. Diese Konfrontation mit zerstörerischer Macht ist auch eine Kon-

hands and are prepared to use them. This confrontation with destructive power is also a confrontation with one's own impotence. Surrounded by the reality of war, identificational play with the role of the armed fighter that some children enact should surprise nobody. The romantic notion that links play with freedom and a lightness of the spirit, with beauty and pleasure, is not applicable here. Richard Schechner, who has researched play and who has also developed an anthropology of theatre, invented the concept of 'dark play' to identify forms of play that are not obviously any such thing. According to Schechner, play does not have to be either voluntary or fun (Schechner 1988: 3). Generally, 'dark play' is used to describe a form of play where the participants are

frontation mit der eigenen Machtlosigkeit. Umgeben von der Wirklichkeit des Krieges, liegt das identifikatorische Spiel mit der Rolle des bewaffneten Kämpfers nahe. Die romantische Vorstellung, wonach Spiel mit Freiheit und Leichtigkeit des Geistes, mit Schönheit und Genuss verbunden ist, greift hier nicht. Der Theateranthropologe und Spielforscher Richard Schechner hat den Begriff des ›Dark Play‹ erfunden, um Spiele zu identifizieren, die sich nicht mehr als solche ausweisen. Spiel, so Schechner, muss weder freiwillig noch lustig sein (Schechner 1988: 3). ›Dark Play‹ bezeichnet im Allgemeinen ein Spiel, bei dem sich die Teilnehmer:innen nicht bewusst sind, dass sie an einem Spiel beteiligt sind oder auch nicht in der Lage sind, ihrer Beteiligung zuzustimmen. ›Dark Play‹ bezieht sich auch auf Spiele, die die üblichen Spielregeln unterlaufen, z.B. Spiele, in denen einige Teilnehmenden leiden, woraus andere ihren Spielspaß beziehen. »Dark Play unterläuft die Ordnung, löst den Rahmen auf und bricht seine eigenen Regeln – so sehr, dass das Spiel selbst in Gefahr ist, zerstört zu werden« (Schechner 2013: 119). Kriegsspiele unter den Bedingungen des realen Krieges sind ebensolche ›dunkle Spiele‹. Teilnehmende und (wir) Außenstehende(n) wissen nicht mehr zu unterscheiden, ob sich die Akteur:innen im Krieg befinden oder ob die kriegerische Wirklichkeit spielerisch aufgehoben ist. Die Gegenüberstellung von Spiel und Ernst macht hier keinen Sinn mehr, und es drängt sich die Frage auf, ob und wie Kinder im Krieg tatsächlich Krieg ›spielen‹ können.

In diesem Beitrag stehen Spiel und Spielzeug im Mittelpunkt und damit die Erlebniswelten von Heranwachsenden. Von Kindern im Kontext von Flucht und Vertrei-

not conscious that they are participating in a game or are not capable of communicating whether they wish to participate or not. 'Dark play' is also used to characterise games that undermine normal rules of games and include, for example, games in which some participants draw their pleasure from playing on the suffering of other participants. "Dark play subverts order, dissolves frames, and breaks its own rules – so much so that the playing itself is in danger of being destroyed" (Schechner 2013: 119). Enacting war games while real wars are raging around the actors should be seen as a form of 'dark play'. Neither the participants nor we external observers, can determine whether the actors are really in a war or whether martial reality has (temporarily) been suspended through playful means. The dichotomy between play and earnestness becomes meaningless, and the question arises if and how children living through a war can really be said to 'play' at warfare.

This article has focussed on play and toys and the worlds that young people experience through these mediums. Children are only normally talked about in the context of flight and expulsion when aid organisations are appealing for funds, when politicians use emotional, hackneyed phrases to drum up support for their positions in difficult times or when the TV news broadcasts a few seconds of vivid footage. All too rarely access is facilitated to the perspectives and experiences that children have themselves. Children's voices are not heard. Examining things that children integrate into their play opens up new per-

bung ist zumeist nur dann die Rede, wenn es um Appelle von Hilfsorganisationen geht, wenn Politiker:innen in schwierigen Zeiten mit emotional aufgeladenen Floskeln für ihre Position werben oder wenn für wenige Sekunden eindringliche Berichte in den Fernsehnachrichten zu sehen sind. Sichtweisen und Erfahrungen von Kindern selbst sind allzu selten wahrnehmbar. Kinder bleiben stumm. Der Blick auf Dinge, die dem Spiel dienen, eröffnet Perspektiven auf die Schutzbedürftigkeit und die Ängste von Kindern, auf ihre Sehnsüchte, aber auch auf Momente der Freude und Ausgelassenheit.

spectives on children's need for protection and their fears, but it also enables insights into children's desires and into their joyful and carefree moments.

Literatur | References

Ariès, Philippe. 1975. *Die Geschichte der Kindheit*. München.

Bosch, Aida. 2014. Identität und Dinge. In: Samida, Stefanie, Manfred Eggert, Hans-Peter Hahn: *Handbuch Materielle Kultur*, 70–77. Stuttgart.

Canby, Sheila R. 1997. Drachen. In: John Cherry: *Fabeltiere. Von Drachen, Einhörnern und anderen mythischen Wesen*, 19–67. Stuttgart.

Csíkszentmihályi, Mihály. 2000. *Das flow-Erlebnis. Jenseits von Angst und Langeweile: im Tun aufgehen*. Stuttgart.

Cockrill, Pauline. 1994. *Die große Enzyklopädie der Teddybären. Portraits von über 500 Teddybären von 1902 bis heute. Das umfassende Nachschlagewerk für alle Liebhaber und Sammler*. München.

Davis, Jac T.M.; Melissa Hines. 2020. How Large Are Gender Differences in Toy Preferences? A Systematic Review and Meta-Analysis of Toy Preference Research. *Archives of Sexual Behavior*, 49: 373–394. https://doi.org/10.1007/s10508-019-01624-7

Dieckmann, Christoph. 2021. »Ich wollte immer meine Kindheit verstehen«. Wolfgang Kohlhaase ist Deutschlands bedeutendster Drehbuchautor. Ein Besuch zum 90. Geburtstag, *Die Zeit* 11/2021, 11. März 2021, 49.

Eigen, Manfred; Winkler, Ruthild.1985. *Das Spiel – Naturgesetze steuern den Zufall*. München.

Falkenberg, Karin. 2015. Notspielzeug. Die Phantasie der Nachkriegszeit. In: Karin Falkenberg: *Notspielzeug. Die Phantasie der Nachkriegszeit. Bürgerausstellung des Spielzeugmuseums, 26. Juni 2015 bis 1. Februar 2016*, 10–11. Petersberg.

Greim, Liane. 2004. *Faszination von Waffen auf Kinder*. München: LMU Hausarbeit zur Zulassung zum 1. Staatsexamen für das Lehramt an Grundschulen. (http://www.schleibinger.com/waffe0/waffe0.pdf).

Huizinga, Johan. 1938. *Homo ludens – Vom Ursprung der Kultur im Spiel*. Hamburg.

Koenneritz, Marie von. 1947. *Margarete Steiff und der Teddybär. Die Geschichte eines Lebens*. Stuttgart.

Landgraf, Claudine; Rosemarie Pfirschke. 2005. *Unterwegs mit Koffer und Teddybär: Europas Kinder und der Zweite Weltkrieg*. Rheinbach.

Paus-Hasebrink, Ingrid. 2010. Die neuen Helden der Kinder und Jugendlichen – Ein Blick in die Rezeptionsforschung. In: Schinkel, Eckhard: *Die Helden-Maschine. Zur Aktualität und Tradition von Heldenbildern*, 60–68, Essen.

Schechner, Richard. 1988. Playing. *Play and Culture* 1: 3–19.

Schechner, Richard. 2013. *Performance Studies: An Introduction*. New York.

Schiller, Friedrich. 1795. Ueber die ästhetische Erziehung des Menschen. [2. Teil; 10. bis 16. Brief.] In: Friedrich Schiller: *Die Horen*, Band 1, 2. Stück, 51–94.

Schmelz, Bernd; Rüdiger Vossen. 1995. *Auf Drachenspuren. Ein Buch zum Drachenprojekt des Hamburgischen Museums für Völkerkunde*. Bonn.

Seidel, Dieter. 2006. *Good Bye, Teddybär. Erinnerungen an eine Kindheit in Krieg und Frieden*. Paderborn.

Talalay, Lauren E. 1993. *Deities, Dolls, and Devices. Neolithic Figurines from Franchthi Cave, Greece*. Indianapolis.

Anmerkungen

1 World Vision: »Spielzeug-Set«, https://www.worldvision.de/spenden/das-gute-geschenk/spielzeugset, abgerufen am 2. November 2021.

2 Der Journalist Nick Robins-Early hat 2015 eine Reihe von beeindruckenden historischen Fotografien zusammengestellt, die spielende Kinder unter den Bedingungen von Krieg zeigen. Siehe Nick Robins-Early: »17 Haunting Historical Photos of Children at Play During Wartime«, *Huffington Post*, 24. März 2015 (https://www.huffpost.com/entry/children-war-photos_n_6904678).

3 Cathrin Hennicke: »Children contemplating suicide in Greece's Moria refugee camp«, *Deutsche Welle*, 21. September 2018 (https://www.dw.com/en/children-contemplating-suicide-in-greeces-moria-refugee-camp/a-45597294).

4 Die Ankleide- bzw. Modepuppe Barbie, die ab den späten 1960er Jahren ihren Siegeszug als junge Erwachsene antritt, repräsentiert eine vergleichsweise neue Entwicklung.

Notes

1 "Spielzeug-Set", World Vision, https://www.worldvision.de/spenden/das-gute-geschenk/spielzugset, Accessed 2 November 2021.

2 In 2015, the photojournalist Nick Robins-Early produced an impressive series of what will go down as historical photos, showing children playing under wartime conditions. See: Nick Robins-Early, "17 Haunting Historical Photos of Children at Play During Wartime", *Huffington Post*, 24 March, 2015. https://www.huffpost.com/entry/children-war-photos_n_6904678

3 Cathrin Hennicke, "Children contemplating suicide in Greece's Moria refugee camp", *Deutsche Welle*, 21 September 2018. https://www.dw.com/en/children-contemplating-suicide-in-greeces-moria-refugee-camp/a-45597294

4 Barbie dolls, which children can dress and undress in various fashionable outfits, came to prominence with the appearance of a young adult in the late 1960s. This, however, was a relatively new development.

5 Both German and Swedish use essentially a single word for dragon and kite, *Drache* or *Drachen* in German, and *Drake* in Swedish, investing a charged, mythic potency in the toy for adults and children.

Schuhe
Drei Schritte eines langen Weges

Shoes
Three Steps in a Long Journey

Anoush Masoudi

> Die Opfer haben häufig keine Fingerabdrücke, da ihre Körper zu lange im Wasser waren ... Es scheint, dass Bilder in den Sozialen Medien und die Kleidung die wichtigsten Hinweise sind, an denen man sich orientieren kann. (M'charek und Casartelli 2019: 739)

> The victims often have no fingerprints because their bodies have been in water too long ... It seems that pictures on social media and the clothing are the most valuable signs to go by. (M'charek and Casartelli 2019: 739)

Wenn die Körper keine Auskunft mehr über die Person und ihre Wege geben, dann wird die Kleidung zum Zeugen, zum Beleg und zur Spur für die Menschen auf der Flucht.[1] Die Darstellung von Geflüchteten hat sich in den vergangenen Jahren durch Internet und Smartphones gewandelt. Handybilder zeigen die Realitäten der Flucht aus Sicht der einzelnen Personen, GPS verzeichnet ihre Routen und Videos dokumentieren besondere Momente aus dem Kontinuum schier endloser Wege. Doch die Bilder und Daten spiegeln nie den ganzen Prozess des Fliehens und Flüchtens. Ein Blick auf die materielle Beschaffenheit der Kleidung ermöglicht hier einen anderen, weiteren Zugang. Mein Augenmerk richtet sich dabei auf die Schuhe.

Schuhe zeigen unterschiedliche Momente, Nöte, Schrammen und Narben, innen wie

When bodies can no longer tell us anything about the people who lived in them or the routes they travelled on, then their clothes become witness, evidence and trace of people who are fleeing.[1] The internet and smartphones have transformed representations of refugees in recent years. Smartphone photos show the reality of flight from individual perspectives, refugees use GPS to plot their routes, and videos document special moments in the continuum of sheer endless journeys. But these photos and data never reflect the whole process of fleeing and of flight. In this context, looking at the materiality of clothing facilitates a different and further-reaching approach. My own particular focus is on shoes.

Shoes reflect different moments, wants,

außen. Im Laufe der Flucht verändern sie ihre Form. Sie sind dabei aufs Engste mit den Körpern der Geflüchteten und zugleich mit den Routen ihrer Flucht in Kontakt. Sie dienen zuallererst dem Schutz der Füße auf dem Weg. Sie schützen die Füße der Flüchtenden, helfen ihnen, Grenzen zu überschreiten, Tag und Nacht zu gehen. Ob auf dem Meer oder in der Wüste, ob es kalt oder heiß ist – am Ende des Weges werden die Schuhe davon sichtbar Zeugnis geben.

Wie können wir mit Blick auf das unscheinbare Alltagsobjekt der Schuhe etwas darüber herausfinden, was mit den Körpern der Flüchtenden geschieht? Wie können wir die Spuren lesen, die die Schwierigkeiten des Weges auf den Schuhen hinterlassen, und wie können sich diese als Symbole für die Gefährdungen des migrantischen Körpers deuten lassen? Dieser Text führt in drei Abschnitten auf eine imaginäre Reise: von der syrisch-türkischen Grenze bis zu jener zwischen Serbien und Kroatien und weiter nach Frankreich, in den ›Dschungel von Calais‹. Die Bilder dreier Fotografen lenken den Blick auf die Schuhe, ihre Materialität, die ihnen eingeschriebenen Geschichten, unsichtbare Wunden, Leid, Ängste und Hoffnungen. Die Fotografen – Kemal Vural Tarlan, Marko Risovic und Mehdi Chebil – berichteten mir in E-Mails davon, wie, wo, wann und warum sie die Fotos aufgenommen haben. Auszüge aus diesen Konversationen zitiere ich auf den folgenden Seiten.

scratches and scars, inside and out. In the course of a journey of flight they change their form. They remain in the closest contact with refugees' bodies and with the routes refugees flee along during this process. They protect the refugees' feet, help them to cross borders and assist them in moving on, day and night. Whether in the sea or in the desert desert, whether cold or hot – at the journey's end, the shoes bear visible witness of where they have been.

How can we scrutinise shoes as an apparently unspectacular everyday object to find out what happens to refugees' bodies when they flee? How can we read the traces marked on shoes by the difficulties of the routes taken? And how can we interpret shoes as symbols for the dangers the migrant's body has to face? Through three sections, the following article leads us on an imaginary journey: from the Syrian-Turkish border, to the border between Serbia and Croatia, and then to France and the 'Calais Jungle'. The images captured by the three photographers direct our gaze towards the shoes, their materiality and the stories engraved on them, reflecting invisible wounds, suffering, anxieties and hopes. The photographers Kemal Vural Tarlan, Marko Risovic and Mehdi Chebil sent me email reports on how, where, when and why they took individual photos. I cite excerpts from these conversations in the following pages.

Erste Etappe: Unterwegs mit ›eisernen Schuhen‹ | Syrisch-türkische Grenze, 2012

In vielen Märchen stoßen wir immer wieder auf Geschichten von ›eisernen‹ oder ›magischen Schuhen‹. Meist handeln sie von einer einzelnen Frau, die sich auf den Weg macht, um ihre Ziele zu erreichen. Geschildert werden die Strapazen und Entbehrungen einer langen Reise, die nötigen Vorbereitungen und allerlei Hürden auf dem Weg.

Die Heldinnen brauchen sieben eiserne Schuhe, um die Strecke zu Fuß zu bewältigen und das verheißungsvolle Ziel zu erreichen. Die richtigen Schuhe spielen dabei eine entscheidende Rolle (Axford 2015). Auch Menschen auf der Flucht brauchen ›eiserne Schuhe‹, um ans Ziel zu kommen, und auch sie müssen Hindernisse bewältigen, Wüsten, Meere und Berge überwinden, um einen sicheren Ort zu erreichen. Im Märchen stehen die eisernen oder magischen Schuhe symbolisch für eine gute Vorbereitung. In der erzwungenen Migration kommt den Schuhen dagegen eine ganz praktische, gleichwohl nicht weniger essenzielle Rolle zu: Ohne sie ist es schlicht unmöglich, heiße Wüsten zu durchwandern oder eine stürmische See und schneebedeckte Berge zu überqueren.

Gute, strapazierfähige Schuhe schon zu Beginn zu verlieren, ist für Flüchtende verhängnisvoll, denn ihr Weg ist noch lang. Auf dem ersten Bild von Kemal Vural Tarlan[2] sehen wir solche verlorenen Schuhe. Der Fotograf dokumentiert seit über 10 Jahren liegengebliebene Dinge an der Grenze zwischen Syrien und der Türkei. Über die verlorenen Schuhe schreibt er:

Phase One: On the Road with 'Iron Shoes' | Syrian-Turkish Border, 2012

In many different fairy-tales, we stumble invariably on stories about 'iron' or 'magic' shoes. They mostly recount the story of an individual woman, who sets off on the road in order to reach the destinations she has chosen. The trials and deprivations faced on a long journey are described, as are the necessary preparations and all manner of obstacles that must be overcome. The female heroes of such tales require seven iron shoes, to successfully manage the route on foot and to reach the apparently auspicious destination. Having the right kind of shoes plays a crucial part in meeting this challenge (Axford 2015). People who are fleeing need the equivalent of 'iron shoes' to reach their destination, as they, too, must overcome hurdles, and traverse deserts, seas and mountains to reach safe places. In fairy tales, the iron or magic shoes symbolise good preparations. In forced migration, shoes take on an entirely practical role, which does not make them less essential; without them it would be simply impossible to cross hot deserts, sail over stormy seas or climb over snow-capped mountains.

Losing good-quality, heavy-duty shoes at or towards the start of their journey is disastrous for refugees, who have a long route ahead of them. On the first photo taken by Kemal Vural Tarlan,[2] we see just such lost shoes. For more than 10 years, the photographer documented things left behind in the border region between Syria

Diese Fotos wurden zwischen 2011–2014 in den verminten Gebieten an der Grenze zwischen der Türkei und Syrien gemacht. Etwa drei Millionen Menschen haben diese verminte Grenze in den drei Jahren überquert. Die Schuhe von Frauen und Kindern blieben zurück, von einem auf den anderen Tag. Am Anfang nahm ich sie mit und steckte sie in meine Tasche …

Eines Tages dann stieß ich nahe der Grenze auf drei einzelne Schuhe. Sie lagen dort in der glühenden Hitze, als würden sie auf ihre Partner warten … Nach diesem Tag nahm ich die Schuhe nicht mehr mit. Ich mache nun diese Bilder.

Wie Tarlan beschrieb, überquerten Anfang 2012 Tausende Menschen die Grenze. Das erste Hindernis, auf das sie stoßen (neben dem nötigen Geld für die Schleuser), ist der Stacheldraht, der sich ihnen physisch in den Weg stellt. Die Grenze, an der das Foto von den Ballerinas entstand, wird nicht von Militär oder Polizei bewacht, ist aber mit Stacheldraht befestigt, der sich gegen die Körper der Flüchtenden richtet. Die Menschen zerschneiden das feindselige Material, springen darauf und darüber, stets mit dem Risiko, sich zu verletzen. Feste Schuhe schützen die Füße vor Schnitten und Wunden. Im Bild sind die Grenzbefestigungen unsichtbar, die Fotos betrachten nur die Schuhe. Das Paar wirkt fast ästhetisch arrangiert, aufeinanderliegend, als sei es eigens für das Foto so drapiert. Die erste Frage, die beim Betrachten in den Kopf kommt: Warum trägt jemand diese Schuhe auf einer so gefährlichen Reise? Flüchtende haben meist keine Zeit oder Gelegenheit, sich vor

and Turkey. This is what he wrote about the lost shoes:

These photos were taken between 2011–2014, in mined areas in the Turkey-Syria border region. About three million people crossed this mined border during this 3-year period. The women's and children's shoes left behind remained there ever since. [On first seeing them], the first thing I did was to pick them up and throw them in my bag …

Then one day I came across three single shoes, near the border. They were standing there under the blazing sun, as if waiting for their spouses … From that day forth, I didn't pick the shoes up. I just took these pictures.

As Tarlan described, thousands of people crossed the border at the start of 2012. The first major barrier such people come across, alongside the obstacle of getting enough money together for the smugglers, is the barbed wire stretched out physically across their path. The border at which the photo of the ballerinas was taken is not guarded by soldiers or police, but it is reinforced by barbed wire, a barrier directed against refugees' bodies. People cut through the antagonistic material and jump on and over it, at constant risk of being injured. Strong shoes at this juncture protect their feet against cuts and wounds. The border defences are not visible on the image; the photograph focuses exclusively on the shoes. The couple give the impres-

Abb. 1: Dunkelblaue Ballerinas, gefunden in den verminten Gebieten an der syrisch-türkischen Grenze, 2013. ©Kemal Vural Tarlan

Image 1: Dark-blue ballerinas, found in the mine-strewn Syrian-Turkish border region, 2013. ©Kemal Vural Tarlan

dem Aufbruch geeignete Schuhe zu kaufen. Menschen, die vor dem Krieg fliehen, nehmen das, was sie zur Hand haben, und tragen es, oft ohne zu wissen, was sie auf dem Weg erwartet. Die Schuhe im Bild haben nicht einmal Schnürsenkel, um sie an den Füßen zu halten.

Ein anderer Blick auf die Schuhe in Abbildung 1: Wir sehen ein Paar dunkelblaue Damenschuhe, zurückgelassen an der syrisch-türkischen Grenze. Ins Auge fällt ihr elegantes Design mit einer dunkelblauen Schleife. Für eine lange Flucht sind sie gänzlich ungeeignet. Wir stellen uns vor, wie die Besitzerin damit zu Hochzeiten und anderen Festen ging. Oder schlicht durch die Straßen ihrer Stadt flanierte. Auf den Krieg ist sie nicht vorbereitet. Dann der Moment der Flucht:

sion of almost having been aesthetically arranged for the shot, lying on top of one another, as if this was material draped specifically so for the photo. The first question that runs through a viewer's head is: Why would anybody wear such shoes on such a dangerous journey? Refugees have mostly no time or opportunities before starting out to buy themselves suitable shoes. People fleeing from war take whatever comes to hand and often carry it without knowing what to expect on the way. The shoes on the photo don't even have shoelaces, with which to keep them on someone's feet.

Let's take another look at the shoes in Image 1: we now see a pair of dark-blue women's shoes, left behind on the Syr-

Du entscheidest nicht von langer Hand, du versuchst, dein Leben zu retten. In diesem Moment bleibt keine Zeit, viele Sachen zu sammeln, passende Schuhe zu finden oder zu kaufen. Diese Schuhe sind gemacht für Zeiten des Friedens, nicht des Krieges. Hier versagen sie ihren Dienst, bieten ihrer Trägerin keinen Schutz und keinen Halt auf dem schweren Weg. Nach einer Weile bleiben sie zurück.

Bis dorthin hat das Gehen durch unwegsames Gelände zwischen Syrien und der Türkei seine Spuren und Narben auf den Schuhen hinterlassen. Im Foto sehen wir die Farbe des Bodens, der langsam das Material der Schuhe angreift und deren Blau seinem ian-Turkish border. Their elegant design with a dark-blue bow catches the eye. They are utterly unsuitable for a long journey of flight. We imagine how their owner went in them to weddings and other celebrations. Or simply sauntered through the streets of her home town. She is not prepared for war. Then the moment of flight: deciding to save your life is not something you do long in advance of the event. But in this moment, there's no time to gather lots of things together or to find or buy appropriate shoes. These shoes are made for times of peace, not for days of war. This is the point at which they fail to serve or protect their wearer and can offer no support on the tough journey. After a while, they get left behind.

Until that sartorial turning point is reached, the trek through almost impassable territory between Syria and Turkey leaves its traces and scars on the shoes. In the photo we see the colour of the soil, that insidiously attacks the shoe's material, slowly transforming the blue into a brown. The shoes are still more or less intact, and they could still be useful, but the process of their deformation has begun. The inner sole is stretched and is detaching itself from the shoe, a sign of the large number of kilometres already covered. It is perhaps also an indication about refugees' bodies during the first stage of their flight: The

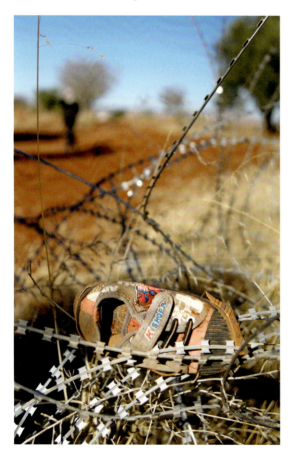

Abb. 2: Kinderschuh, im Grenzzaun verfangen, 2013.
©Kemal Vural Tarlan

Image 2: A children's shoe stuck in a border fence, 2013.
©Kemal Vural Tarlan

Schuhe

Braun anverwandelt. Die Schuhe sind noch weitgehend intakt und könnten noch nützlich sein, doch der Prozess ihrer Verformung hat begonnen. Die Innensohle ist geweitet und löst sich vom Schuh, ein Zeichen des langen Weges. Vielleicht auch ein Hinweis auf die Körper der Flüchtenden auf dieser ersten Etappe: Die Füße werden müde und beginnen zu schmerzen. Sie verkrampfen beim Versuch, die untauglichen Schuhe am Fuß zu halten, gegen den Widerstand des Untergrunds, in den sie sinken. Sie beginnen zu brennen, während die Hitze des Bodens durch die dünne Sohle dringt.

Im zweiten Bild von Kemal Tarlan sehen wir einen Kinderschuh, gefangen im Stacheldraht einer Grenzbefestigung. Knapp ein Drittel der über 80 Millionen Geflüchteten weltweit (2020) sind Kinder unter 12 Jahren. Um die 300.000 Kinder werden jedes Jahr auf der Flucht gar geboren (UNHCR 2021: 3). So lauten die registrierten Zahlen des Weltflüchtlingshilfswerks UNHCR. Wie viele Kinder nie einen sicheren Ort erreichen und unterwegs zu Tode kommen, lässt sich kaum sagen. Zwischen 2014 und 2018 wurden nach Angaben der UN 1.600 Kinder auf der Flucht für tot oder vermisst erklärt (Laczko et al. 2019: 1). Doch viele bleiben ungezählt. Das Bild des zweijährigen Alan Kurdi, der auf der Flucht über das Mittelmeer ertrank, ging 2015 um die Welt. Die Situation für Kinder auf der Flucht hat sich seither nicht gebessert.

Das Foto hier zeigt uns den Schuh eines vier- bis sechsjährigen Kindes. Das Kind selbst ist nicht zu sehen und vermutlich bereits an einem anderen Ort, doch das Bild symbolisiert eindrücklich die Gefahren erzwungener Migration: Die scharfen Klingen des Stacheldrahts umfangen den Schuh und scheinen ihn bald zu verfeet get tired and get cramps in efforts to keep the unapt shoes on, working against the resistance of the route's surface, into which they are sinking. And the feet begin to burn, as the heat from the ground penetrates through the thin soles.

In Kemal Vural Tarlan's second photo, we see a children's shoe stuck in barbed wire, a part of a series of border defences. Almost a third of a total of over 80 million refugees worldwide (2020) are refugees under the age of 12, and around 300,000 children every year are born while their parents are fleeing – according to numbers registered with the UNHCR, which holds a mandate to aid and protect refugees.[3] It is almost impossible to say with any accuracy how many refugee children never reach a safe location and die en route. UN figures state that 1600 refugee children were registered as dead or missing during the 2014–2018 period (Laczko et al. 2019: 1). But many remain uncounted. In 2015, the picture went global of two-year old Alan Kurdi, who drowned in the Mediterranean while his family was attempting to flee to Europe. Despite such repeated tragedies, the situation for refugee children has not improved.

Kemal Tarlan's photo shows us the shoe of a child aged between four and six. The child itself is nowhere to be seen, and has presumably moved on long ago, but the image is impressive in its symbolisation of forced migration: The wire's sharp points have surrounded the shoe and will apparently soon devour it. For a while, this shoe protected the child's foot, but now it has been left behind for good. How is the child meant to continue on their journey?

schlingen. Eine Zeitlang hat der Schuh die Füße eines Kindes geschützt, doch nun bleibt er auf dem Weg zurück. Wie soll das Kind seinen Weg fortsetzen? Es bräuchte einen eisernen Schuh, doch nun hat es vielleicht gar keinen mehr. Tarlan schreibt über das Bild:

> Die Kinderschuhe standen auf diesen Drähten. Ich glaube, es waren die ersten Monate von 2012. Tausende Menschen überquerten die Grenze jeden Tag. Schleuser:innen arbeiteten überall an der Grenze, offen und ohne Angst vor irgendjemandem. Wenn du genau hinschaust, kannst du sogar die Silhouette eines Schleusers im Hintergrund des Fotos sehen ...

Der Schuh im Bild ist von Wind und Wetter staubig und brüchig geworden. Können wir uns vorstellen, was mit dem Körper eines verletzten Kindes geschieht? Ein Kind, das mit einer Wunde am Fuß weiterlaufen muss, in der Hitze und Kälte, auf hartem Boden für lange Zeit, ohne passende Schuhe. Die Wunde wird tiefer, schmutzig, entzündet sich, wird lebensbedrohlich. Was aus dem geflüchteten Kind geworden ist, wissen wir nicht. Wir sehen den Schuh: beschädigt, außen wie innen.

Zweite Etappe: High Heels im Schlamm und Plastiktüten-Schuhe | Serbisch-kroatische Grenze, 2015

Es war ein kalter Tag im späten Oktober 2015, irgendwo in der Vojvodina-Ebene, nahe der Grenze zwischen Serbien und Kroatien. Die

They actually need 'iron shoes', but now, perhaps, they have no shoes at all.
This is how Tarlan describes the image:

> The children's shoe was balanced in the wire. I think it was during the first months of 2012. Thousands of people were crossing the border every day. Human traffickers were working all along the border, openly, not scared of anyone. If you look carefully, you can see a human trafficker's silhouette in the background ...

The shoe on the photo has been made dusty and brittle by the wind and the weather. Can we imagine what these same conditions could do to the body of an injured child? A child with a wounded foot, who must walk on, through the cold and the heat, on hard ground for a long, long time, without suitable shoes? The wound grows deeper, gets dirty and starts to become life-threatening. We do not know what happened to this child who had to flee. All we see is the shoe: damaged both inside and out.

Phase Two: High Heels in the Mud, and Plastic Bag Shoes | The Serbian-Croatian Border, 2015

It was a cold day in late October 2015, somewhere on the Vojvodina Plain, near the border between Serbia and Croatia. A foggy morning started to emerge from the darkness of the night, and shapes of human figures began to appear out of

Abb. 3: Modische Schuhe auf schlammigem Boden an der serbisch-kroatischen Grenze, 2015. ©Marko Risovic

Image 3: Fashionable shoes on muddy ground at the Serbian-Croatian border, 2015. ©Marko Risovic

Dunkelheit der Nacht ging gerade in einen nebeligen Morgen über, und die Umrisse menschlicher Gestalten tauchten aus dem Nebel auf. In Decken gehüllt, sichtbar und offensichtlich frierend (was leicht an ihrer Körpersprache erkennbar war, selbst in der morgendlichen Dunkelheit), versuchten die Menschen sich an den gerade entzündeten Feuern zu wärmen, an denen sich immer mehr Leute sammelten. Die Umrisse der Menschen zeichneten sich gegen die flackernden Flammen ab, die die Rettung vor dem Erfrieren bedeuteten. Bald war die Nacht vergessen, aber die matschige Ebene war voller Neuankömmlinge, so weit das Auge reichte.

Marko Risovic,[3] ein serbischer Dokumentarfotograf, fotografiert seit Jahren Geflüchtete als ›Unerwünschte‹ auf der Balkanroute. Er beschreibt eine kalte Nacht, in der Geflüchtete die Grenze zwischen Serbien und Kroatien

the fog. Wrapped in blankets, obviously and visibly freezing – as could easily be recognised in their body language, even in the morning darkness – they tried to warm themselves around the newly-lit fires, where larger and larger groups had begun to gather. The people's silhouettes could be glimpsed in the flickering flames that were saving them from freezing to death. The night was soon forgotten, but the muddy plain was full of people arriving as far as the eye could see.

Marko Risovic, a Serbian photojournalist, has been photographing refugees – who are seen by many locals as "undesirables" – on the so-called Balkanroute. He describes a cold night during which refugees reached the border between Serbia and Croatia, one further phase in their overall journey. Whereas the refugees are

erreichen. Eine weitere Etappe des Weges. Wo die Flüchtenden an der türkischen Grenze auf Stacheldraht und Sonnenhitze stoßen, begegnen ihnen hier Kälte und Schlamm. Wir stellen uns vor, wie die Geflüchteten unserer Geschichte ihren Weg von der Türkei auf der Balkanroute fortgesetzt haben, tagelang, nächtelang. Die Frau und das Kind, die ihre Schuhe an der türkischen Staatsgrenze verloren hatten, haben es bis zur nächsten Staatsgrenze geschafft, mit oder ohne Schuhe. Risovic erinnert sich:

> Ich schaute auf meine eigenen Füße. Meine Schuhe waren komplett mit Matsch bedeckt ... Dann wurde mein Blick auf die Kinderstiefel gezogen, die drohten, im tiefen Matsch des Oktoberregens stecken zu bleiben, hier, 100 Meter von der unsichtbaren Linie, die Länder und Leben voneinander trennt ... Und dann sah ich eine ähnliche Szene, ein alter Mann, der direkt neben mir stand und sich gerade so auf dem rutschigen Boden über einem Grenzgraben halten konnte ... Dann entdeckte ich einen weiteren matschigen Schuh und schließlich sah ich Schuhe, die nach einer Nacht in diesen Zuständen gerade poliert worden waren, als ob der Besitzer sie für eine Festlichkeit vorbereiten wollte.

Risovic beschreibt die sehr unterschiedlichen Beziehungen, die die Menschen an der Grenze zu ihren Schuhen haben. Manche kommen in einfachen Schlappen, andere machen Schuhe aus allen möglichen verfügbaren Materialien, wieder andere sorgen sich um die Sauberkeit ihrer Schuhe. Der starke Kontrast zwischen

most likely to come across barbed-wire and hot sunshine at the Turkish border, here they are met with cold and mud. We could choose to imagine how the refugees in our story have continued from Turkey on the Balkan Route, for days and nights on end. The woman and the child, who lost its shoes at the Turkish border, have made it to the next boundary between nation-states, with or without shoes.

This is what Risovic remembers:

> I looked at my feet. My shoes were completely covered in mud ... Then my attention was drawn to children's boots that were in danger of remaining trapped in the deep mud of the October rains, only 100 metres from the invisible line that separates countries and lives ... And then I saw a similar scene of an older man standing right next to me, barely holding on to a slippery surface above a border trench ... I then spotted another muddy shoe, before catching sight of shoes that, after a night in these conditions, looked like they had just been polished, as if their owner had prepared them for some kind of festivities.

Risovic describes the abundantly different relationships people at the border have with their shoes. Some arrive clad in simple flip-flops, others have made or make shoes out of all possible materials they can access, while others still are concerned about their shoes staying clean. The strong contrast between polished shoes and the muddy ground during the freezing-cold

den polierten Schuhen und dem schlammigen Untergrund in einer eiskalten Nacht verweist auf soziale Verhältnisse im Spiegel der Dinge. Während die einen nicht einmal Schuhe haben, pflegen andere deren Erscheinung. Warum sollte sich jemand in dieser Extremsituation um das Aussehen seiner Schuhe kümmern? Welche Rolle spielen Schönheit und Sauberkeit unter diesen Bedingungen?

Ein Blick auf das erste Bild von Marko Risovic mag Aufschluss geben. Wir sehen die Füße einer Frau mit modischen Schuhen auf schlammigem Untergrund. Das Bild gibt die Situation der Flucht wieder: falsche Schuhe, falsche Körper am falschen Ort. Die Schuhe wirken zu groß, scheinen der Frau nicht recht zu passen. Doch die Füße sind nichtsdestotrotz in Bewegung. Das Bild betont so die Rolle des Gehens und die Schwierigkeiten der Fortbewegung.

Risovic zeigt bewusst nicht das Gesicht der Frau, sondern richtet die Kamera auf die Füße und Schuhe. Im Blick auf den schlammigen Boden erzeugt er statt Mitgefühl einen starken Kontrast zwischen dem modischen Schuh und dem unwirtlichen Untergrund, zwischen Schönheit und Dreck, zwischen Mode und Wirklichkeit, zwischen Elend und Würde.

Im Bild sind die Knie der Frau nicht zu sehen, doch die Stellung der Beine legt vorsichtiges Gehen nahe, der Körper passt sich den Gegebenheiten an. Die Stiefel sind nicht sauber, doch verleihen sie der Trägerin den Rest ihres eigenen individuellen Stils. Oder sind Schönheit und Mode auf der Flucht nicht von Belang?

Wenn diese Frau die Grenze im Angesicht der Polizei überqueren muss, wird sie damit nicht schnell laufen können, um sich zu retten. Im Schutz einer größeren Gruppe von

night highlights the social relations discernible in the reflections that various objects generate. While some don't have any shoes at all, others are still taking care of what their shoes look like. Why would someone in this extreme situation be bothered about their shoes' appearance? What role is being played by beauty and cleanliness under these conditions?

Marko Risovic' first photo could give answers to these questions. We see the feet of a women, dressed in fashionable shoes, muddy shoes. The image reflects the larger situation of fleeing: the wrong shoes, the wrong bodies at the wrong places. The shoes seem too large and don't appear to fit the woman properly. Despite this, the feet are in movement. The picture underscores the role of walking and the difficulties involved in making progress.

Risovic chooses consciously to omit the woman's face and focus on the feet and the shoes instead. In directing our attention towards the muddy ground, he generates, instead of empathy, a strong contrast between the trendy shoes and the inhospitable ground they walk upon, between beauty and dirt, between fashion and reality and between misery and dignity.

The woman's knees are not visible in the picture, but the position of the legs suggests that she is walking carefully, the body adapting to the conditions. The boots are not clean, but they may grant their wearer what remains of an individual style. Or are we to believe that beauty and fashion are irrelevant to people who are fleeing?

When this woman has to cross the border under the watchful gaze of the police,

Geflüchteten wird sie langsamer gehen können und so eine größere Chance haben, die Grenze zu überwinden. Der Fotograf Risovic beschreibt eine solche Situation:

> Sie huschten durch die Felder zum informellen Grenzübergang, voller Hoffnung, dass sie heute weiterkommen und die nächste eisigkalte Nacht näher an ihrem Zielort verbringen würden. Eine Menschenmenge drängte sich bereits vor der Polizeiabsperrung. Sie riefen: »Öffnet die Grenze!« Eine Gruppe aus Alt und Jung, aus Männern und Frauen und Familien mit vielen Kindern.

Im zweiten Bild von Risovic sehen wir einen Mann ohne Schuhe, aber mit etwas wie

she won't be able to run fast in order to save herself. Protected, by contrast, amongst a larger group of refugees, she will be able to walk slower and thus have a greater chance of being able to cross the border. Risovic describes a situation like this from a photographer's perspective:

> They scurried through the fields to the informal border crossing, full of hope that they would be able to proceed that same day, and that they would spend the next icy night somewhere that was at least closer to their final destination. The crowd was already squeezed in front of a police cordon made up of many officers. They shouted: "Open the border!" A mass composed of young

Abb. 4: Schuhe aus Plastiktüten an der serbisch-kroatischen Grenze, 2015. ©Marko Risovic

Image 4: Shoes made from plastic bags at the Serbian-Croatian border, 2015. ©Marko Risovic

Schuhe

Schuhen, notdürftig gefertigt aus Plastiktüten. Ohne viel zu zeigen, verdeutlicht das Foto die Härten der Flucht. Hier geht es nicht mehr um falsche Schuhe am falschen Ort, sondern die Aneignung von ›falschem‹ Material für etwas wie Schuhe. Die Plastiktüten sind Müll, Abfall anderer Leute. Ihren ursprünglichen Nutzen und Wert haben sie verloren, sie sind als Unrat nicht mehr für den Gebrauch bestimmt. Das Material wird gleichsam transformiert, und das Objekt, das entsteht, entzieht sich der Bezeichnung: Sind das überhaupt Schuhe?

Der Mann hat seine Füße zunächst in Alufolie gewickelt und sie dann mit weißen Plastiktüten umhüllt. Schuhe sind das nicht; die Knappheit und Not erzeugen ein neues Ding. Das Gehen damit fällt schwer, die Schutzfunktion des Schuhwerks füllt es nur unzulänglich aus. Weder Kälte noch spitze Steine hält es ausreichend ab. Das neue Ding imitiert die Eigenschaften der fehlenden Schuhe und bleibt doch notgedrungen überall dahinter zurück. Es ist ein Unding, nur etwas mehr als nichts.

Wenn sich die Grenze öffnet und eine neue Etappe bevorsteht, wird der Mann nicht weit kommen. Er wird genötigt sein, sein notdürftiges Schuhwerk zu wechseln. Denn die Tüten werden reißen, abfallen und wieder Abfall werden. Doch in seiner Erinnerung wird er die Schuhe aus Plastiktüten wohl weiter tragen, als Erinnerung an Not und Schmerz, Kränkung und Leid. Er wird sie nicht los auf dem endlosen Weg, ohne Schuhe, mit Müll an den Füßen. Plastiktüten mit einer deutschen Flagge darauf.

Gefragt, warum er als Fotograf überhaupt diese Aufnahmen der Schuhe macht, resümiert Marko Risovic:

and old, men and women, and families with lots of children.

In the second picture taken by Risovic, we see a man without shoes, who is wearing something *like* shoes nonetheless, objects made from plastic bags. Without revealing much, these materialities expose the hardships of flight. This is no longer about wearing the wrong shoes at the wrong place, but rather about adapting the 'wrong' material to make something like shoes. The plastic bags are rubbish, objects other people have thrown away. They have lost their original use and value, and, as garbage, they seemed unlikely to be used in any way again. While this material is transformed with the change of use, the object that emerges resists categorisation: Are these shoes?

The man in the photo has first wrapped his feet in aluminium foil and then enveloped these tucked-in body parts in white plastic bags. Shoes are not what these things are: The paucity of the materials and the urgency of the situation create a new thing, or rather two new things. Walking using these things is hard going, as they are unable to fulfil the protective function of real footwear. They are incapable of sufficiently keeping out sharp stones or the cold. These new 'things' imitate the characteristics of the missing shoes yet fall short of meeting what we expect from shoes, in all aspects. They are a nonthing, only slightly more than nothing.

When the border opens, and a new stretch of the route opens out before the refugees, the man on the photo won't make it far wearing his bag-shoes. He will have no option but to change his emer-

Ich fotografierte diese Schuhe. Sie hatten etwas merkwürdig Symbolisches an sich. Sie erzählten gewissermaßen mehr von der Geschichte als nur von den Körpern, die sich in einer Menschenmenge vor der Absperrung drängen. Sie erzählten von der langen Zeit, die mit dem Laufen und dem Warten, der Angst und der Hoffnung verbracht wird. Sie zeigten Spuren von der syrischen Wüste, dem Kopfsteinpflaster in der Türkei, Meersalz und Sand aus Griechenland, mazedonischen Wiesen und serbischen Wäldern, bis sie irgendwann in der Nähe Kroatiens im Matsch stecken blieben. Die Schuhe trugen die Spuren von Geschichte und menschlichem Leiden jener, die auf der Suche nach Freiheit, Frieden und Menschlichkeit die Härten des Lebens durchlaufen.

Dritte Etappe: Endlich Europa, verloren und zukunftslos | Frankreich 2018–2020

Stellen wir uns vor, dass die Polizei die Grenze zwischen Serbien und Kroatien nach einer Zeit öffnet und Geflüchtete passieren lässt. Sie setzen ihren Weg nach Westeuropa fort, mit hochhackigen Stiefeln oder Plastiktüten an den Füßen. Manche bekommen neue Schuhe, etwa von Wohltätigkeitsorganisationen, andere laufen in ihren alten, wieder andere ohne Schuhe. Nach Tagen und Nächten unterwegs erreichen sie Frankreich. Viele erhoffen sich hier das Ende ihrer Reise, ersehnen einen sicheren Ort und einen freundlichen Empfang. Andere werden versuchen, ihre Reise fortzusetzen und den Ärmelkanal zu überqueren, um Großbritannien

gency footwear. The bags will tear, will fall off, and will become rubbish again. But he will continue to wear these shoes made from plastic bags in his memories, recollections constituted from despair and pain and insult and suffering. He won't be able to get rid of them on his endless journey, without shoes, but with rubbish on his feet. Plastic bags with a German flag on them.

Asked why capturing these images of the shoes was important to his work as a photographer, Marko Risovic responded:

> I photographed those shoes. They carried some strange symbolism in them, and, in a way, they told many more stories than the human bodies huddled in a crowd in front of a nervous police formation. The shoes talked about the long time [they had] spent walking, standing, waiting, fearing and hoping … They carried traces of the Syrian desert and Turkish cobblestones, of the sea salt and Greek beaches, of Macedonian meadows and Serbian forests, until they finally got stuck somewhere in the mud near Croatian territory. The shoes bore the traces of history and of human suffering and had been worn by those who had broken through the hardships of life, in search of freedom, peace and humanity.

Phase Three: Europe at Last, Lost and Without a Future | France, 2018–2020

Let us imagine that, after some time, the police open the border between Serbia

Abb. 5: Ein Paar schwarze Schuhe, gefunden in den Puythouck Wäldern nahe Grande-Synthe, 2019. ©Mehdi Chebil

Image 5: A pair of black shoes, found in the Puythouck Woods near the town of Grande-Synthe, in the Nord-Pas de Calais region, France, 2019. ©Mehdi Chebil

zu erreichen. Doch die Wahrheit sieht anders aus: Auch hier sind sie nicht willkommen.

Erahnen lässt sich dies in einem ersten Bild des französischen Fotografen Mehdi Chebil[4] aus einem Wäldchen östlich von Calais. Zu sehen ist ein Paar schwarzer, am Boden liegender Schuhe. Der Hintergrund mit den nassen bunten Blättern deutet auf eine regnerische Nacht im Herbst hin. Wir haben Schuhe im Stacheldraht an der syrisch-türkischen Grenze gesehen und Schuhe im Schlamm an der serbisch-kroatischen. Doch was hat es mit diesen Schuhen, verloren in einem französischen Wald, auf sich?

Mehdi Chebil, der seit langem das Leben von Geflüchteten in Frankreich, nicht zuletzt im so genannten ›Dschungel von Calais‹[5], fotografisch dokumentiert, gibt eine Antwort:

and Croatia and let the refugees through. They continue on their journey to Western Europe, wearing high-heeled boots or plastic bags on their feet. Some get new shoes, perhaps from charity organisations, while others walk on in their old footwear – with a third group wearing no shoes at all. After many days and nights on the road, they finally reach France. Many hope that this will be the end of their journey and long for a safe place to be and a friendly reception. Others will attempt to continue their journey by crossing The Channel to reach the UK. But the truth of the matter is entirely the opposite: They aren't welcome here either.

Such a narrative can be deduced from the first photo reproduced here by the French photographer Mehdi Chebil,[4] taken in a small wood east of Calais. We see a pair

Das Foto mit der Nahaufnahme der schwarzen Schuhe habe ich am 9. Dezember 2018 in den Puythouck-Wäldern nahe Grande-Synthe (25 Kilometer östlich von Calais) aufgenommen. Die Polizei hatte ein paar Tage vorher ein kleines Lager von Migrant:innen zerlegt. Ich machte dieses Foto, weil es meiner Meinung nach zeigt, wie Migrant:innen weglaufen müssen, wenn die Polizei anrückt, und zwar in einer solchen Eile, dass sie wertvolle Gegenstände, wie z. B. Schuhe, zurücklassen.

Die Widrigkeiten sind hier nicht materiell wie der Schlamm oder Stacheldraht. Die Polizei attackiert die Geflüchteten alle 48 Stunden und entwendet sämtliche Habseligkeiten. Hier ist niemand willkommen. Einmal mehr müssen die Menschen weglaufen, dieses Mal vor der Polizei. Ihre Sachen bleiben zurück. Vielfach wird davon berichtet, dass Beamte gerade Schuhe und Schnürsenkel mitnehmen, um die Fortbewegung zu erschweren.[6] Denn sie wissen natürlich, dass es sich ohne Schuhe kaum laufen lässt.

Das Foto verweist auf das Schicksal eines Geflüchteten, der die Schuhe wahrscheinlich auf der Flucht vor der Polizei verloren hat. Wir können uns vorstellen, wie der Mann sich gefühlt hat, der, wie andere, Tage und Wochen unterwegs gewesen war. Mit Schuhen, getragen auf dem beschwerlichen Weg. Schuhen, gepflegt und poliert. Schuhen mit Absätzen, Schuhen aus Abfall, hin zu einem vermeintlich sicheren Ort. Und dann der Angriff durch die Polizei, der alles zerstört, was sie noch besitzen.

Auf ambivalente Weise konterkariert

of black shoes lying on the ground. The background, with its wet, colourful leaves, suggests a rainy night in autumn. We've already seen shoes caught in barbed-wire at the Syrian-Turkish border and shoes stuck in the mud at the Serbian-Croatian border. But what do these shoes, lost in French woods, have to say for themselves?

One answer is provided by Mehdi Chebil, experienced in documenting the lives of refugees in France using photography, including, not the least, lives in the socalled 'Calais Jungle':[5]

The picture with the close-up of the black shoes was taken on 9 December 2018 in the Puythouck woods near Grande-Synthe, 25 kilometres east of Calais. The police had dismantled a small migrant camp in Puythouck a few days earlier. I took this picture because it illustrates, I argue, how migrants have to rush away when the police move in, the haste so dominant that they leave behind precious belongings such as shoes.

The primary adversities in this context are not directly material, in the way that mud or barbed wire is. For as long as migrant camps remained in this area, the police attack refugees every 48 hours or so and steal their belongings. This is a place where no-one is welcome. Yet, people have to run away again, this time from the police. Their things get left behind. It is reported frequently that police officers consciously aim at taking refugees' shoes and shoelaces off them, to hinder their forward movement.[6]

Schuhe

wird dieser Akt der Wegnahme und Zerstörung von Habseligkeiten durch die nahezu gleichzeitige Schenkung von Kleidung durch Wohltätigkeitsorganisationen, die Chebil beschreibt:

> Ich hatte mit Hilfsorganisationen gesprochen, die Migrant:innen unterstützen und die mir die Kleiderspenden zeigten, insbesondere auch einen Haufen von Schuhen. Ich hatte gesagt, dass es so aussehen würde, als hätten sie genügend Schuhe für alle Migrant:innen. Sie hatten mir erklärt, dass dies nicht der Fall sei, da ihnen die geläufigste Größe, die 42, fehle. Als ich diese Schuhe in den Puythouck-Wäldern sah, erinnerten sie mich an dieses Gespräch und ich fotografierte

The police are naturally aware that the refugees can hardly move without shoes.

The photo highlights what happened to a single refugee, who probably lost his shoes while fleeing from the police. We can imagine how the man felt about this loss, after having been on the road for days and weeks like so many others with the shoes worn on this arduous journey. Shoes that had been looked after and polished. Shoes with heels, shoes made from rubbish, shoes bearing people to what they supposed would be a safe place. And then the attack by the police, destroying everything they still possessed.

In an ambivalent fashion, this act of forced removal and destruction of possessions is countered by the gifting of clothing

Abb. 6: Schenkung von Schuhen durch Wohltätigkeitsorganisationen, 2019. ©Mehdi Chebil

Image 6: Shoes donated to charitable organisations, 2019. ©Mehdi Chebil

sie, weil sie aussahen, als hätten sie die Größe 42.

Mit Chebil lässt sich fragen, was die Logik der Situation ist, wenn eine Gruppe Schuhe an Flüchtende ausgibt und die Polizei an gleicher Stelle Schuhe und andere Habe entwendet.

Zu Abbildung 7 berichtet der Fotograf Mehdi Chebil:

> Hamed, ein iranischer Migrant, hatte keine Schuhe mehr, nachdem die Polizei ihm sein Zelt weggenommen hatte. Seitdem das große Lager »der Dschungel« aufgelöst wurde, haben die französischen Autoritäten eine Taktik der Schikane eingeführt, nach der sie die kleinen Lager alle 48 Stunden auseinandernehmen. Am frühen Morgen kommt ein großes Aufgebot an Polizei und nimmt alles weg, was sich auf dem Gebiet des Lagers befindet – Zelte, Kleidung, Schuhe und

by charities, which often happens almost simultaneous to such police attacks, as described by Chebil:

> I had earlier talked to charities that assist migrants, which showed me the large mounds of donations they received, and most specifically a pile of shoes for migrants. I told them that it looked like they had enough shoes for all migrants [who wanted them] but they answered: "No, we don't," because they were actually missing the most common shoe-size, 42. When I saw these shoes in the Puythouck woods, it reminded me of that discussion, and then I shot the photo, because it looked like the shoes were about size 42.

Chebil's observations can be used to inquire into the logic of a situation, in which one group hands out shoes to refugees, while, at the same point in the refugees' lives, the police steal their shoes and their other belongings.

The photographer Mehdi Chebil has the following to say about Image 7:

> Hamed, an Iranian migrant, has been left with no shoes after the police has taken his tent away from him. Since clearing the large Calais Jungle camp, the French authorities have had a policy of harassing migrants by disman-

Abb. 7: Ohne festes Schuhwerk, 2019. ©Mehdi Chebil
Image 7: Without stable footwear, 2019. ©Mehdi Chebil

Essen. Die Migrant:innen haben sich angepasst, sie ziehen mit ihrem Zelt und ihren Habseligkeiten entlang der Straße, zehn oder zwanzig Meter entfernt vom ehemaligen Lager. Aber manchmal vergessen sie etwas und können es nicht mehr holen, sobald die Polizei angekommen ist. Das war Hamed, dem Mann im Bild, passiert.

Es ist ein Moment voll Traurigkeit. Das Gesicht von Hamed sehen wir nicht, doch der Blick auf seine Füße zeugt von Elend und Einsamkeit. Die kurze Beschreibung von Chebil informiert uns, dass es eigentlich auch ein Zelt mit weiteren Dingen gab, die nun nicht mehr da sind. Das Bild zeigt Dinge, ohne sie zu zeigen; die abwesenden Gegenstände sind präsenter als jene, die wir sehen. Wir sehen die Füße von Hamed, aber keine Schuhe.

Es wird kalt; deshalb das kleine Feuer, das selbst einen verlorenen Eindruck macht. Es ist nicht stark genug, um Hamed lange zu wärmen, ohne Zelt und ohne Schuhe. Er sollte lieber gehen; es ist nichts da, das ihn hier hält. Das Bild verweist auf Fehlendes:

Zum einen auf die Abwesenheit des Individuellen. Hamed könnte irgendjemand sein. Er ist ein Stellvertreter all der Verlorenen auf der Flucht. All jener, die den Weg nicht antreten oder bewältigen konnten. Das Kind, das seinen Schuh an der syrisch-türkischen Grenze verlor. Die Frau mit den blauen Ballerinas und jene mit den Stiefeln im Schlamm. Der Mann mit den Plastiktüten an der serbisch-kroatischen Grenze. Und Hamed selbst, ein in Frankreich Gestrandeter.

tling their little camps every 48 hours or so. Numerous French police arrive early in the morning and take away everything – tent, clothes, shoes or food – that has been left on the patch of land where the camp was. Migrants have adapted and simply move their tent and belongings to the roadside, [sometimes only] 10 or 20 meters from the [previous] camp. But sometimes they forget something and can't retrieve it once the police has arrived. This is what happened to Hamed, the Iranian man shown on the picture.

The photo gives access to an excerpt out of the life of a refugee who has come to the provisional end of his journey. It documents total loss, a moment in which it is not clear how things can go on at all. These people carry heavy experiences with them and are exhausted, both physically and mentally. While en route, with painful feet and defective shoes, they kept hold of a dream of Western Europe as a safe place. And now here they are, and they are everything but safe.

It is a moment full of sadness. We do not see Hamed's face, but the view of his feet bears witness to misery and loneliness. The short description that Chebil provides informs us that, until that point in time, a tent had actually existed containing other things that are now no longer there. The photograph shows things without showing them: The absent objects are more present than any we actually see. We see Hamed's feet, but we do not see any proper shoes.

It's getting cold, that's why a small fire can be seen on the photo, a combustion

Shoes

Zum zweiten repräsentiert das Bild die Abwesenheit von Hoffnung. Das Grau in Grau, das erlöschende Feuer, die ungeschützten Füße. Hamed macht keine Schuhe aus Müll; er scheint paralysiert und verloren inmitten eines fremden Waldes.

Drittens, sehen wir die Abwesenheit von Schuhen. In der Kälte wären sie bitter nötig, doch die Reise muss ohne sie weitergehen, wenn überhaupt, auf der Suche nach einem sicheren Ort.

Viertens schließlich erkennen wir die anwesende Abwesenheit der Staatsgewalt. Wir wissen aus der kurzen Beschreibung von Mehdi Chebil, dass die Polizei da war, da ist, und den Geflüchteten ihre Sache wegnimmt. Doch im Bild ist sie nicht zu sehen. Sie bedingt und prägt die Situation und entzieht sich zugleich ihrer Lokalisierung. Zurück bleibt Hamed, ein Verlorener am Ende seines Weges.

Du bist in Europa, doch ohne deine Schuhe. Für viele Geflüchtete sind neue Schuhe, sobald sie angekommen sind, von großer Bedeutung: zum einen, um die Härten und die dunklen Momente der Flucht zu vergessen, die sich als Spuren auf den Schuhen abgezeichnet haben; zum anderen, um in den Sozialen Medien zu zeigen, dass sie es geschafft haben, selbst wenn das erreichte Ziel nichts mit ihren ursprünglichen Vorstellungen zu tun hat. Wenn Hamed oder andere Geflüchtete also mit letzter Energie weitergehen, wenn sie nicht krank oder von der Polizei aufgehalten und abgeschoben werden, wenn sie irgendwo ankommen, in einem Flüchtlingscamp in Deutschland etwa, kaufen sie sobald wie möglich neue Schuhe. Sie machen Fotos mit diesen neuen Schuhen, um den Schmerz ihrer Körper und ihrer Erinnerungen zu mil-

that itself gives the impression of being lost. It cannot suffice to warm Hamed for long, when he no longer has a tent or shoes. He should go: There is nothing there to keep him here any longer. The image points towards what is missing.

First, anything individual is absent: Hamed, at least as photographed here, could be anyone. He has become a representative of all those who are lost, who are fleeing. And of all those who do not even start out on the journey or who were unable to cope with the burdens it produced. The child who lost its shoe at the Syrian-Turkish border. The women wearing the blue ballerinas and the woman wearing the high heels in the mud. The man with the plastic bags on his feet at the Serbian-Croatian border. And Hamed himself, stranded in France.

Second, the image represents the absence of hope. The grey inside the grey, the fire that is petering out, the unprotected shoes. Hamed does not make new shoes out of rubbish. He appears to be paralysed and lost in the middle of a foreign forest.

Third, we see the absence of shoes. They would be bitterly necessary in this cold. But if Hamed's journey is to go on at all, it must go on without them, in the search for a safe place.

Fourth, we ultimately recognise the present absence of state authority. We know from Mehdi Chebil's short description that the police had been there, is there, and takes the refugees' things away from them. But the police are nowhere to be seen on the image. They are the force that conditions and moulds the situation

dern, was aber kaum gelingt. Sie leben weiter mit den Narben ihrer Flucht.

Denn für diese Reise wären selbst ›eiserne Schuhe‹ nicht genug. Dieses Märchen von den Schuhen der Menschen auf der Flucht kennt kein Happy End.

but that, simultaneously, avoids being localised by it. It is Hamed who remains behind, a lost person at the end of his journey.

You are in Europe – but without your shoes. New shoes acquire great significance for many refugees on arriving: for some, in order to forget the privations and dark moments of their journey of flight, which left its traces on their shoes; for others, in order to show on social media that they have made it, even if the destination arrived at has nothing in common with their original notions. So, if Hamed or other refugees manage to move on with the last energy they have inside them, if they manage not to get ill or stopped by the police and deported, and if they finally arrive somewhere, at a refugee camp in Germany, for example, they buy themselves new shoes as soon as they can. They take photos of themselves wearing these new shoes, to mitigate the pains in their bodies and the pain of their memories, though this will hardly be a comprehensive mitigation. They will live on with the psychological and physical scars that fleeing has brought them.

Because, for this journey, even 'iron shoes' could not suffice. This (fairy) tale of refugees' shoes cannot be given closure with a happy ending.

Literatur | References

Axford, Martin. 2015. Various: In Their Shoes: Fairy Tales and Folktales. *School Librarian* 63 (4): 234.

Laczko, Frank; Julia Black; Ann Singleton. 2019. »Fatal Journeys Volume 4. Missing Migrant Children«. *International Organization for Migration (IOM)*. (https://publications.iom.int/system/files/pdf/fatal_journeys_4.pdf).

M'charek, Amande; Sara Casartelli. 2019. Identifying Dead Migrants: Forensic Care Work and Relational Citizenship. *Citizenship Studies*, 23(7): 738–757.

UNHCR. 2021. *Global Trends – Forces Displacement in 2020*. https://www.unhcr.org/60b638e37.pdf

Anmerkungen | Notes

1. Von diesem Umstand erzählt auch das Interview mit Margherita Bettoni im Beitrag »Die letzten Dinge«: https://materialitaet-migration.de/objekte/die-letzten-dinge/
2. Mehr zu diesem Dokumentarfotografen unter: http://www.kemalvuraltarlan.net/
3. Mehr zu diesem Dokumentarfotografen unter: https://kamerades.com/member/marko-risovic/
4. Mehr zu diesem Fotografen unter: https://www.chebilink.com/
5. Als ›Dschungel von Calais‹ wurde eine Zeltstadt mit provisorischen Unterkünften nahe der französischen Stadt Calais bezeichnet, die 2015 entstand und im Oktober 2016 von der Polizei geräumt wurde. Vgl. u.a. Michel Agier: *Der »Dschungel von Calais«. Über das Leben in einem Flüchtlingslager*, Bielefeld 2020.
6. Maliheh Bayat Tork: »Schnürsenkel, Gürtel, GPS, Google Maps: Die wesentlichen Dinge Flüchtender«, Materialität der Migration, abgerufen am 9.12.21, https://materialitaet-migration.de/en/objekte/shoe-lace-belt-gps-google-map-vital-things-to-the-people-on-the-run/

1. "Die letzten Dinge", Accessed art: https://materialitaet-migration.de/objekte/die-letzten-dinge/
2. For more on this photo-journalist, see the following, accessed 16 November 2021: http://www.kemalvuraltarlan.net/
3. UNHCR, "Global Trends – Forces Displacement in 2020", p. 3. Accessed on 16 November 2021. https://www.unhcr.org/60b638e37.pdf
4. For more on this photographer, see the following, accessed 16 November 2021: https://www.chebilink.com/
5. The Calais Jungle was the name given to the town of tents and other makeshift dwellings which grew up near the French town of Calais in 2015, from which the police evicted the inhabitants in October 2016. Cf. Michel Agier et al., *The Jungle. Calais's Camps and Migrants*, trans. David Fernbach (Cambridge, UK: Polity Press, 2019).
6. Maliheh Bayat Tork: Shoe lace, Belt, GPS, Google Map: Vital Things to the People on the Run. Accessed 16 November 2021. https://materialitaet-migration.de/en/objekte/shoe-lace-belt-gps-google-map-vital-things-to-the-people-on-the-run/

Menstruationsprodukte

Menstruation Products

Maliheh Bayat Tork | Antonie Fuhse

Abb. 1: Pakete mit Binden für die Verteilung an Migrant:innen, Maschhad, Iran 2017. ©Maliheh Bayat Tork

Image 1: Packages of sanitary pads for distribution to migrants, Mashhad, Iran 2017. ©Maliheh Bayat Tork

Auf dem obigen Bild sehen wir Kartons mit Binden, Einwegrasierern, Seife und Shampoo, die für die Verteilung an Migrant:innen in der iranischen Stadt Maschhad bestimmt sind. Die Binden werden von den Hilfsorganisationen vor Ort an Frauen und Mädchen verteilt, eine Maßnahme, die unter das so genannte ›Menstrual Hygiene Management‹ fällt, das mittlerweile ein wichtiger Bereich der humanitären Hilfe darstellt.

Nachfolgend beschäftigen wir uns mit

The photo shows boxes of sanitary pads, disposable razors, soap and shampoo intended for distribution to migrants in the Iranian city of Mashhad. The sanitary pads were distributed to women and girls by local aid organisations in what has become a central aspect of humanitarian aid: 'Menstrual Hygiene Management'.

The following discusses the question of how menstruating people[1] deal with their menstruation during flight and mi-

der Frage, wie Menstruierende[1] während der Flucht und Migration mit ihrer Menstruation umgehen. Den Zugang zu diesem Thema bilden die Materialitäten der Menstruation wie das Blut, der Körper und die Menstruationsprodukte. Dabei verstehen wir Menstruation immer auch als eine Perspektive auf andere Themen: auf Gesundheit, auf Geschlechtergerechtigkeit, auf das Frausein sowie auf die gegenderte Erfahrung von Flucht, Migration und Ankommen.

Das Bild aus Maschhad und unsere Beschäftigung mit dem Thema Menstruation und Migration bewegte uns als Autorinnen des Textes dazu, unsere eigenen Menstruationserfahrungen zu reflektieren. Dabei kamen wir schnell auf die vielen Umschreibungen zu sprechen, die benutzt werden, um nicht direkt von der Menstruation zu reden. Und so stießen wir darauf, dass die »rote Tante« nicht nur in Deutschland, sondern auch im Iran einmal im Monat zu Besuch kommt (اومده خاله قرمزی).

Menstruation – überall gleich?

Wir beide sind in etwa 5.000 Kilometer Entfernung voneinander aufgewachsen – Maliheh in Maschhad im Iran und Antonie in Gotha, Deutschland. Als wir uns das erste Mal trafen, um über Menstruation und Migration zu sprechen, stellten wir allerdings bald fest, dass wir ähnliche Erfahrungen gemacht hatten: Sowohl in Deutschland als auch im Iran wird kaum offen über Menstruation gesprochen. Stattdessen wird die Periode umschrieben mit »Ich habe meine Tage« oder »meine Erdbeer-

gration. We approach these questions via the materialities of menstruation, such as blood, the body and menstrual products. In doing, so we understand menstruation as a lens through which to address other issues: health, gender equality, being a woman and the gendered experience of flight, migration and arrival.

The picture taken in Mashhad and our preoccupation with the issue of menstruation and migration moved us, as the authors of this text, to reflect on our own menstrual experiences. We quickly got on to discussing the many euphemisms people use to avoid talking directly about menstruation. We found out, for example, that the "red aunt" (German = *rote Tante*; Iranian = خاله قرمزی اومده) comes to visit once a month not only in Germany, but also in Iran.

Is Menstruation the Same Everywhere?

Maliheh grew up in Mashhad, Iran, about 5,000 kilometres away from where I grew up in Gotha, Germany. That said, when we first met to talk about menstruation and migration, we soon realised that we had had very similar experiences. Both in Germany and in Iran, menstruation is rarely talked about openly. Instead, the period is described as "that time of the month" or, to translate directly from the German euphemisms: "I have my days" (*ich habe meine Tage*), "my strawberry week" (*meine Erdbeerwoche*) or indeed a visit by the "red aunt". The fact that this latter euphemism is used in both German and Iranian nation-

Abb. 2: Menstruate with Pride by Sarah Maple, 2010–11.
©Sarah Maple, https://www.sarahmaple.com/paintings-3

Image 2: Menstruate with Pride by Sarah Maple, 2010–11.
©Sarah Maple, https://www.sarahmaple.com/paintings-3

woche« und eben mit der »roten Tante«, die zu Besuch kommt. Dass diese Umschreibung in beiden Kontexten verwendet wird, macht zwei Aspekte deutlich: Bei der Menstruation handelt es sich um eine fundamentale Realität, die einen Großteil der Weltbevölkerung betrifft, über die aber kaum direkt und offen gesprochen wird und die häufig mit Scham behaftet ist.

Aktivist:innen, Künstler:innen und Frauenrechtsbewegungen treten bereits seit Jahrzehnten dafür ein, die Menstruation von negativen Assoziationen und Praktiken zu befreien. Trotzdem wird die Menstruation im

al contexts points to two central aspects: Menstruation is a fundamental reality that affects a large part of the world's population but which is rarely spoken about directly or openly – and which is often fraught with shame.

Activists, artists and women's rights movements have been campaigning for decades against negative associations and practices identified with menstruation. Despite this, menstruation is often portrayed in the dominant public conversation as something impure and polluting that has to be hidden with the help

dominanten Diskurs häufig als etwas Unreines und Verschmutzendes dargestellt, das mit Hilfe von ›Hygieneprodukten‹ versteckt werden muss. Trotz dieser Gemeinsamkeiten in Erfahrung und Umgang mit der Periode, gibt es kulturspezifische Praktiken und Normen. In Ländern wie Benin und Kamerun wird die erste Menstruation von Feierlichkeiten begleitet; in anderen Kontexten werden Frauen während ihrer Periode vom Besuch religiöser Stätten ausgeschlossen (siehe Hawkey, Ussher und Perz 2020: 107). Die Erfahrungen, die Menstruierende machen, werden somit sowohl vom Kontext beeinflusst als auch von der individuellen Geschichte geprägt und verändern sich während des Lebens (siehe Hawkey et al. 2017: 1473–1490).

Materialitäten der Menstruation und Migration

Menstruierende auf der Flucht setzen sich mit den gleichen Dingen auseinander wie Menstruierende generell. Allerdings verschärfen Flucht und Migration bestimmte Aspekte wie die Versorgung mit Menstruationsprodukten und Unterwäsche, den Zugang zu sanitären Anlagen und Waschmöglichkeiten sowie die Entsorgung von Menstruationsprodukten. Die Schamgefühle, die häufig mit der Menstruation verbunden werden, beeinflussen auch die Interaktionen mit anderen Geflüchteten oder dem Personal in Unterkünften und Camps (siehe Tellier et al. 2020: 594–595).

Unsere Beschäftigung mit den Materialitäten der Migration beginnen wir mit den Infrastrukturen, die sich zu diesem Thema im Bereich der humanitären Hilfe entwickelt ha-

of 'hygiene products'. Despite this commonality in experience and the handling of the period, many practices and norms are culture-specific. In countries like Benin and Cameroon, a girls first menstruation is celebrated; in other contexts and regions, women are excluded from visiting religious sites during their period (cf. Hawkey, Ussher and Perz 2020: 99–113). The experiences of people who menstruate are thus influenced by their specific cultural contexts as well as by their individual biographies and life paths (cf. Hawkey et al. 2017: 1473–1490).

Materialities of Menstruation and Migration

Menstruating people in the process of flight grapple with the same issues as menstruators in general. That said, flight and migration exacerbate certain aspects of menstruation, such as access to menstrual products and underwear, access to sanitation and washing facilities as well as for the disposal of menstrual products. The feelings of shame often associated with menstruation also influence interactions with other refugees or staff in refugee housing facilities and camps (see Tellier et al. 2020: 594–595).

We begin our discussion of the materialities of migration with the infrastructure that has been developed regarding menstruation in the field of humanitarian aid. In the context of flight and migration, aid organisations have a major influence on people's access to a wide variety of re-

ben. Im Kontext von Flucht und Migration haben Hilfsorganisationen einen großen Einfluss auf den Zugang der Menschen zu den verschiedensten Ressourcen, wie auch zu Menstruationsprodukten und sanitären Anlagen.

Das ›Menstrual Hygiene Management‹

Die Bezeichnung ›Menstrual Hygiene Management‹ (MHM) zeigt, dass im humanitären Bereich der Umgang mit der Menstruation als eine Frage von Hygiene angesehen wird. Diese Eingruppierung wird durchaus kritisiert und alternative Begriffe wie ›Menstrual Health and Management‹ genutzt, um den Fokus auf die Gesundheit zu legen und weniger auf den Umgang mit etwas vermeintlich ›Dreckigem‹ (Tellier et al. 2020: 594).

Das Thema Menstruation wurde erst relativ spät als wichtiger Bereich der humanitären Hilfe erkannt. Ein erster Schritt in diese Richtung wurde im Jahr 2000 gegangen, als die UNFPA (Bevölkerungsfonds der Vereinten Nationen) die Empfehlung aussprach, so genannte ›dignity kits‹ mit Menstruationsartikeln bereitzustellen. Erst durch diese Empfehlung waren Hilfsorganisationen dazu berechtigt, die Finanzierung solcher Maßnahmen zu beantragen. Der Fokus wurde in den folgenden Jahren erweitert, bis auch die Bereitstellung von sanitären Anlagen mit sauberem Wasser sowie von Privatsphäre und Sicherheit unter die förderfähigen Maßnahmen aufgenommen wurden (Tellier et al. 2020: 597). Die konkreten Maßnahmen des MHM unterscheiden sich je nach lokalem Kontext und werden unter anderem in dem ausführlichen sources, including to menstrual products and sanitary facilities.

'Menstrual Hygiene Management'

The term 'Menstrual Hygiene Management' (MHM) reveals that menstruation is framed as a question of hygiene in the field of humanitarian aid. Many see this framing critically, and alternative terms such as 'Menstrual Health and Management' are also used, to shift the focus towards health issues and away from implications that one is dealing with something that is supposedly 'dirty' (Tellier et al. 2020: 594).

Only relatively recently was the issue of menstruation recognised as an important sphere of humanitarian aid. A first step in this direction was taken in 2000, when the UNFPA (United Nations Population Fund) recommended distributing so-called 'dignity kits' that include menstrual products. This recommendation was the precondition for aid organisations to be able to apply for funding for such measures. The focus was then expanded over the years, until the provision of sanitary facilities with clean water as well as privacy and security were also included in these policy actions (Tellier et al. 2020: 597). Specific MHM measures differ depending on the local context and are summarised, for example, in the detailed handbook *A Toolkit for Integrating Menstrual Hygiene Management (MHM) into Humanitarian Response*. The challenges identified in this "toolkit" include: lack of safe and private spaces for dealing with menstruation;

Maßnahmenkatalog A *Toolkit for Integrating Menstrual Hygiene Management (MHM) into Humanitarian Response* zusammengefasst. In diesem ›Werkzeugkasten‹ werden unter anderem folgende Herausforderungen genannt: Mangel an sicheren und privaten Orten für den Umgang mit der Menstruation, Verlegenheit und Angst, Mangel an Informationen, Überbelegung der Lager sowie kulturelle Tabus und Restriktionen. Als Maßnahmen werden das Befragen von Frauen und Mädchen über ihre Erfahrungen und Präferenzen, die Bereitstellung von Informationen über die Menstruation, von Menstruationsprodukten sowie von sicheren, frauenfreundlichen sanitären Anlagen und von angemessenen Entsorgungsmöglichkeiten für Menstruationsprodukte empfohlen (Sommer, Schmitt und Clatworthy 2017).

Aber was ist mit den Personen außerhalb der offiziellen Infrastrukturen von Lagern und Hilfsorganisationen? Welche Erfahrungen machen Menstruierende auf der Reise? Die folgende Geschichte von Selma, die sie im Rahmen des Forschungsprojektes »Zur Materialität von Flucht und Migration« bei einem Interview im Durchgangslager Friedland erzählte, gibt uns einen Einblick.

Der Körper, Migration und Menstruation – Selmas Geschichte

Um zu ihrem Mann zu kommen, der schon eine Weile in Deutschland war, verließ Selma Ende 2019 Al-Hasakah, den kurdischen Teil im Nordosten Syriens, begleitet von vier weiteren Frauen. Eine der Frauen war sechzehn,

overcrowding in camps; shame and fear; lack of information, particularly for menstruating people needing to use facilities; cultural taboos; and other restrictions. One measure proposed for addressing these issues is to talk with women and girls about their experiences and preferences, about menstrual product provision, about safe, female-friendly sanitary facilities and appropriate facilities for the disposal for menstrual products, and, finally, about the provision of information on menstruation (Sommer, Schmitt and Clatworthy 2017).

But what about people outside the official refugee infrastructure and aid organisations? What do menstruating people experience while in transit? The following story about Selma, which she narrated during an interview in the Friedland Transit Camp as part of the research project "On the Materiality of (Forced) Migration", gives us insight.

The Body, Migration and Menstruation – Selma's Story

To join her husband who had already been in Germany for some time, Selma left Al-Hasakah, the Kurdish part of northeastern Syria, in late 2019, accompanied by four other women, one of them 16 and the others in their 20s. Once in Turkey, they stayed briefly in a camp with women from Syria and other countries. Their first attempt to cross the border from Turkey to Greece was as a group of seven women, and without any traffickers to guide them.

die anderen waren in ihren Zwanzigern. In der Türkei angekommen, blieben sie für kurze Zeit in einem Lager mit einigen anderen Frauen aus Syrien und anderen Ländern. Bei ihrem ersten Versuch, die Grenze von der Türkei nach Griechenland zu überqueren, waren sie sieben Frauen und wurden nicht von Schleppern geführt. Es gelang ihnen, an der Grenze anzukommen, wo sie von der griechischen Grenzpolizei aufgegriffen und in einem LKW festgehalten wurden, in dem sich bereits 50 geflüchtete Männer aus Afghanistan befanden. Der LKW brachte sie zur Grenzpolizeistation, wo sie verhört wurden.

Die Polizei beschlagnahmte ihre Mobiltelefone, und die Frauen wurden aufgefordert, sich auszuziehen, um durchsucht zu werden. Als sie sich weigerten, wurden sie von einem der Polizeibeamten als »Huren« bezeichnet. Die Frauen wehrten sich, woraufhin die Polizei zu Gewalt griff und sie schlug. Sie wurden gezwungen, sich auszuziehen. Als sie über dieses Erlebnis sprach, stiegen Selma Tränen in die Augen und sie schluckte einen Kloß im Hals hinunter. Wenn sie davon sprach, was die Polizei ihnen angetan hat, nutzte sie das Wort »Vergewaltigung«.

Die Frauen waren zwei Tage und Nächte lang in Polizeigewahrsam, ohne dass sie Wasser oder Essen bekamen. Sie wurden im selben Raum mit den geflüchteten Männern festgehalten. Wir fragten Selma, ob sie sich im selben Raum mit den Männern sicher fühlten; sie sagte, dass sie von den afghanischen Männern nichts zu befürchten hatten, trotzdem wechselten sie sich beim Schlafen ab, um sich zu schützen. Sie fuhr fort: »Nach zwei Tagen wurden wir mit einem Lastwagen an die Küste gebracht und dann wie Kartoffelsäcke in ein

They made it to the border but were then arrested by the Greek border police and detained in a truck with 50 male refugees from Afghanistan. The truck transferred them to the border police station where they were interrogated.

The police confiscated their mobile phones and ordered the women to undress, so they could be frisked. When they refused, one of the police officers called them "whores". The women defended themselves, whereupon the police resorted to violence and hit them. They were forced to undress. Selma had tears in her eyes and a lump in her throat when she talked about this experience. When she spoke about what the police had done to them, she used the word "rape".

The women were in police custody for two days and two nights, without receiving any food or water. They were kept in the same room as the male refugees. I asked Selma whether they felt safe in the same room with the men. She said that they had no reason to fear the Afghan men, but took turns sleeping nonetheless, to protect themselves. She continued: "After two days, we were taken to the coast in a truck and then dumped into a boat like sacks of potatoes." They arrived somewhere early the next morning, apparently on the Turkish coast, where they were moved onto the shore. They started walking, hoping to reach civilisation. After several hours, they reached a motorway where they saw a police convoy. It was the Turkish police. The women approached them. "They offered us water and food," said Selma with a trembling voice.

Boot verfrachtet«. Am frühen Morgen kamen sie irgendwo an, anscheinend an der Küste der Türkei, und wurden dort ausgesetzt. Sie begannen zu laufen, in der Hoffnung, irgendwo anzukommen. Nach Stunden erreichten sie eine Autobahn, wo sie eine Polizeikarawane sahen. Es war die türkische Polizei. Die Frauen gingen dorthin. »Sie boten uns Wasser und Essen an«, sagte Selma mit zittriger Stimme.

Einen Moment lang zögerte Selma, etwas zu sagen. Sie hielt kurz inne, schaute nach unten, dann hob sie den Kopf, als hätte sie viel Mut fassen müssen, um zu erzählen, dass sie alle, die sieben Frauen, gleichzeitig ihre Periode bekommen hatten, wahrscheinlich aufgrund von Stress und Anspannung. Da sie keine Menstruationsprodukte bei sich hatten, zerrissen sie einige ihrer Hemden und benutzten sie als Binden.

Selmas Geschichte verdeutlicht mehrere Aspekte. Sie zeigt, dass Gender einen Einfluss darauf hat, welche Erfahrungen während Flucht und Migration gemacht werden. Es gibt sicherlich Erfahrungen, die von vielen Menschen auf der Flucht geteilt werden, wie Gefühle von Ohnmacht, Stress, Angst oder die Hoffnung, dass sich die eigene Situation bald bessern wird. Allerdings gibt es Faktoren, die einen großen Einfluss auf die individuellen Erfahrungen haben. Dazu gehören neben Gender auch Alter, Staatsbürgerschaft, Ethnizität, Religionszugehörigkeit, Gesundheit, Sexualität oder Elternschaft.

Deutlich wird anhand der Erzählung von Selma zudem, dass Migration und Flucht verkörperte Erfahrungen sind. Die Körper der Frauen reagierten auf den ständigen Stress mit der Menstruation. Für die Frauen war diese Reaktion wiederum eine zusätzliche Belastung und stellte sie vor die ganz prak-

For a moment, Selma hesitated. She paused briefly, looked down, and then she raised her head, as if she had had to work up lots of courage to narrate that the whole group, all seven women, had all gotten their period simultaneously at this point – probably due to the stress and tension. As they had no menstruation products with them, they tore apart some shirts they had and used the strips as pads.

Selma's story contains several telling aspects. It shows that gender influences experiences of flight and migration. There are certainly experiences that people who are fleeing share: feelings of powerlessness, stress, fear or hope that one's situation will soon improve. However, certain factors also have a major impact on individual experiences. In addition to gender, these include age, citizenship, ethnicity, religious affiliation, health, sexuality and the status of being a parent.

But Selma's story also demonstrates that migration and flight are bodily experiences. The women's bodies in her narrative responded to the constant stress by menstruating. For the female protagonists, this physical reaction was an additional burden and presented them with the very practical challenge of having to somehow catch the blood with the help of strips of fabric torn from their clothing.

Visible and Invisible Blood

We encounter representations of blood in medical context, in adverts to donate blood or in television programmes, where

tische Herausforderung, das Blut irgendwie mit Hilfe von Stoffstreifen, gerissen aus ihrer Kleidung, auffangen zu müssen.

Blut – sichtbares und unsichtbares

Abbildungen von Blut begegnen wir in medizinischen Kontexten, bei der Werbung für Blutspenden oder im Rahmen von Serien oder Filmen, wo es in Kampfszenen mitunter ästhetisch arrangiert wird. In Kirchen sehen wir den an das Kreuz geschlagenen Jesus, Blut an den Hand- und Fußgelenken und am Kopf unter dem Dornenkranz. Menstruationsblut dagegen ist medial so gut wie unsichtbar oder wird negativ dargestellt als etwas Dreckiges und Unhygienisches (Hawkey, Ussher und Perz 2020: 107). Historisch und kulturübergreifend wurde Menstruationsblut in vielen Fällen als giftig, magisch oder verunreinigend konstruiert – als Zeichen des »monströsen Weiblichen« (Ussher 2006: 6). Das Verstecken des Periodenblutes ist die Norm, die auch den Umgang mit dem eigenen Körper während der Menstruation prägt.

Künstler:innen und Aktivist:innen versuchen seit einigen Jahren, dieser Norm des Versteckens entgegenzuwirken und das Menstruationsblut aktiv aufzuwerten, es also entweder in einem positiven oder in einem vieldeutigen Licht zu zeigen (Green-Cole 2020: 787–788). Judy Chicago war eine der Pionier:innen auf

it is sometimes arranged aesthetically in fight scenes. In churches, we see Jesus nailed to the cross, blood on his wrists and ankles, and on his head under the crown of thorns. Menstrual blood, on the other hand, is as good as invisible in the media, or it is portrayed in negative terms as something dirty and unhygienic (Hawkey, Ussher und Perz 2020: 107). Historically and across cultures, menstrual blood is often constructed as poisonous, magical or contaminating – as a sign of the "monstrous feminine" (Usher 2006: 6). Hiding period blood is the norm and defines the way people who menstruate deal with their own bodies during menstruation.

Abb. 3: Perioden-Emoji
Image 3: The period emoji

For several years now, artists and activists have been trying to counteract this norm of hiding menstrual blood, repositioning it to show it in a positive light or at least an "ambiguous" one (Green-Cole 2020: 787–788). Judy Chicago was one of the pioneers in this sphere, known, among other things, for her lithograph *Red Flag* (1971), in which she showed the removal of a used tampon. For many women, this practice is a part of everyday life during their period they are more than familiar with. But the visual representation of this act is virtually nonexistent. Since the representational breakthroughs achieved by Chicago and others, social media has become an important location for thematizing menstruation and for breaking down the taboos associated with men-

diesem Gebiet und wurde unter anderem für ihre Lithografie *Red Flag* (1971) bekannt, auf der sie das Entfernen eines gebrauchten Tampons zeigte. Für viele Frauen ist dies eine Praxis, die zu ihrem Alltag während der Periode gehört, die ihnen vertraut ist, deren visuelle Darstellung aber so gut wie nicht existent ist. Mittlerweile werden die sozialen Medien zu einem wichtigen Mittel, um die Periode zu thematisieren und Menstruationsblut zu enttabuisieren. Beispielsweise gibt es seit 2019 ein offizielles Perioden-Emoji für Messenger-Dienste (siehe Abbildung 3).

Auch auf der Flucht und im Leben im Camp ist es für Menstruierende wichtig, das Menstruationsblut unsichtbar machen zu können, um mögliche Stigmatisierungen zu vermeiden. Dieser Wunsch beeinflusst die Wahl der Menstruationsprodukte. In vielen Camps und Erstaufnahmeeinrichtungen werden Menstruationsprodukte an die Ankommenden verteilt. Sie gehören zu den ersten Dingen, die neben Seife, Shampoo, Zahnbürste und Zahnpasta ausgeteilt werden.[2] Zusätzlich gibt es Rasierschaum oder Windeln, je nach Bedarf der Familien.

Tampons, Binden und Menstruationstassen

Für den Umgang mit der Menstruation gibt es heutzutage eine große Bandbreite von Produkten, von Binden und Tampons bis hin zu Periodenunterwäsche und Menstruationstassen. Die Auswahl und die Nutzung dieser Produkte hängen von unterschiedlichen Faktoren ab.

Maliheh arbeitete im Iran für eine Hilfsorganisation, die sich um die Versorgung von

strual blood. An official period emoji was developed, for messenger services in 2019, shown on Image 3.

While fleeing and in refugee camps, it is important for menstruating people to be able to make menstrual blood invisible, to avoid possible stigmatisation. This desire also influences the choice of menstrual products. In many camps and other refugee accommodation, menstrual products are distributed to new arrivals; they are among the first things to be handed out, together with soap, shampoo, toothbrushes and toothpaste.[2] Shaving foam and disposable nappies are also distributed, depending on the needs of the respective individuals.

Tampons, Sanitary Pads and Cups

A broad range of products exists on the market for dealing with menstruation, from sanitary pads, to period underwear, to menstrual cups. The selection and use of these products depend on various factors.

Maliheh worked for an aid organisation in Iran committed to helping Afghan refugees. She was involved in the distribution of care packages to people who were classified as "needy". She was sometimes directly involved in their physical distribution, and also participated in the evaluation of the distribution procedures. This is how Maliheh describes these experiences:

> During my tenure working for a humanitarian aid organisation in Iran, I was involved in providing aid and

Abb. 4: Inhalt eines Hygienepakets für Migrant:innen in Maschhad, Iran 2017. ©Maliheh Bayat Tork

Image 4: The contents of a hygiene packet for migrants in Mashhad, Iran 2017. ©Maliheh Bayat Tork

afghanischen Geflüchteten kümmerte. Sie war in die Prozesse rund um die Verteilung von Versorgungspaketen an Personen beteiligt, die als »bedürftig« eingestuft wurden. Maliheh war teilweise direkt bei der Verteilung dabei und arbeitete bei der Evaluation nach der Verteilung mit. Sie berichtet von dieser Arbeit Folgendes:

> Gewöhnlich werden Pakete mit Menstruationsprodukten für Frauen und Mädchen sporadisch von unterschiedlichen humanitären Organisationen verteilt. Dazu gehören sowohl internationale als auch lokale Nichtregierungsorganisationen (NGOs). In Zeiten besonderer Krisen, wie Erdbeben oder Überflutungen, wie sie in den vergangenen Jahren im Iran stattfanden, gehören Binden immer zu den Dingen, die verteilt werden. Neben

assistance to Afghan refugees, who had been identified as "needy", either in the distribution phase itself, or in postdistribution evaluation. Traditionally, hygiene packages for women are distributed sporadically by different humanitarian organisations, including international agencies and local NGOs operating in Iran. In crises such as earthquakes and floods, which caused large dislocations in Iran in recent years, simple menstrual pads are always among the items distributed. Occasionally, female doctors and midwifes also raised awareness on the importance of menstrual hygiene after menstruation products had been handed out. During home visits to distribute packages and fill out forms, I chatted fre-

der Verteilung von Menstruationsprodukten informierten von Zeit zu Zeit Ärztinnen oder Hebammen über die Bedeutung der Menstruationshygiene. Während des Verteilens der Versorgungspakete und des Ausfüllens der dazugehörigen Formulare unterhielt ich mich mit den Frauen und Mädchen in jedem Haushalt. Die männlichen Mitglieder meines Teams verließen dabei den Raum. Ich sprach mit meinen iranischen und afghanischen Kolleginnen darüber, wie wir seit unserer Jugend bis heute mit unserer Menstruation umgehen. Dabei stellten wir fest, wie sich alltägliche Praktiken, Traditionen und Vorstellungen über die Zeit hinweg verändert hatten. Einige wenige Kolleginnen meinten, dass eine Frau ihre Füße während ihrer Periode nicht waschen sollte, sie ihren Körper nicht kaltem Wasser aussetzen oder sie nicht kochen sollte. Diese Verhaltensregeln wurden wiederum von anderen stark abgelehnt.

Die folgenden Aussagen stammen aus meiner (Malihehs) Zeit im Iran oder meinen Gesprächen mit einigen afghanischen Frauen, die im Flüchtlingslager Moria lebten. Ich sprach mit ihnen nach dem Ausbruch des zweiten Feuers, als sie mit vielen anderen in Zelten am Straßenrand lebten.

> Amineh, eine 34 Jahre alte Frau aus dem Iran, benutzt Tampons, weil sie diese angenehmer findet als andere Produkte. Mit den Tampons kann sie während ihrer Periode schwimmen gehen, erklärt sie. Allerdings erzählt sie auch, dass sie vorsichtig sein müsse und sie die Tampons

quently with women and girls, with the male members of my team leaving the room during these conversations. I also spoke with my Iranian and Afghan female colleagues about ways with which we've dealt with our menstruation, from our youths until today. We concluded that everyday practices, traditions and notions concerning menstruation had changed over time. A few female colleagues were convinced that a woman should not wash her feet or expose her body to cold water during her period, or thought that she should not cook. Other colleagues vehemently rejected such behavioural rules.

The following notes come directly from my (Malileh's) experiences in Iran, and from my conversations with Afghan women in the Moria Refugee Camp in Greece in 2020, after the second fire had broken out. I talked with the women while they were still living in tents by the roadside, along with many others.

> Amineh, a 34-year-old Iranian women uses tampons; she finds them more comfortable than other menstrual products. She explains that with tampons she can still go swimming when she has her period. However, she does say that she has to be very careful: She hides her tampons in her handbag, because if her mother were to see them, she would discover that Amineh is not a virgin. If her mother were to realise that Amineh has had sex before

in ihrer Handtasche verstecke: Wenn ihre Mutter die Tampons sehen würde, würde sie herausfinden, dass Amineh Sex vor der Ehe hatte und sie großen Ärger bekommen würde.[3]

Masoumeh ist eine 27 Jahre alte Frau, deren Eltern aus Afghanistan in den Iran geflohen sind. Sie selbst wurde im Iran geboren und wuchs dort auf. Sie nutzt am liebsten Stoffstreifen, die sie wäscht und in der Sonne trocknen lässt. Andere Produkte würden ihrer Aussage nach Allergien und Infektionen verursachen.

Golrizeh, eine 26 Jahre alte Frau, die aus Afghanistan in den Iran geflohen ist, freut sich, dass sie Binden bekommt und diese nutzen kann. Sie erklärt: »Diese Binden muss ich nicht waschen, ich werfe sie einfach in den Müll. Sie ersparen mir viel Zeit und Arbeit.«

Menstruationsprodukte werden manchmal unter den bedürftigen afghanischen Familien verteilt. Einige Frauen behalten allerdings die traditionelle Art und Weise bei, mit dem Menstruationsblut umzugehen: Verschlissene Kleidung wird in rechteckige Stücke geschnitten, die als Binden genutzt werden. Sie werden gewaschen und in der Sonne getrocknet, um sie wiederverwenden zu können. Fatima erklärt, warum sie an dieser Praxis festhält: »Ich habe drei Töchter, mit mir sind wir zu viert. Ich kann es mir nicht leisten, jeden Monat Menstruationsbinden für uns vier zu bezahlen. Es ist eine Menge weggeworfenes Geld.«

marriage, Amineh would be in big trouble.[3]

Masoumeh, a 27-year-old women, born and raised in Iran as the child of Afghan refugees, does not use menstruation products, preferring strips of fabric that she then washes and dries in the sun. She states that menstruation products cause allergies and infections.

Golrizeh, a 26-year-old Afghan refugee woman in Iran, is glad to receive and to use menstruation pads. She explains: "With these pads, I don't need to wash anything again. I simply throw them in the trash. They save me a lot of time and labour."

Those Afghan refugee families living in Iran with adult daughters, and also categorised as needy, are also sometimes given menstruation products. But some of the women in these families continue to practice traditional ways of dealing with their menstrual blood: Torn clothes are cut into rectangles, which are then used as sanitary pads or towels. They are then washed and dried in the sun, to be reused. Fatima explains why she keeps to this practice: "I have three daughters, meaning there are four in the family including myself. I cannot afford to buy sanitary towels for us four every month. That would be a huge waste of money."

Havva, an Afghan woman in the Moria Refugee Camp, was very happy to receive menstrual products. She ex-

Menstruation Products

Havva, eine Frau aus Afghanistan, die im Lager Moria lebt, ist glücklich über die Menstruationsprodukte. Sie erklärt: »Sie bringen uns Menstruationsbinden. Das ist wirklich gut. Sie sollen Infektionen von uns fernhalten.« Allerdings gibt sie zu bedenken: »Das ist umsonst, solange der Zustand der Toiletten eine Katastrophe ist.«

Marzieh: Da ihre Menstruation immer noch ausbleibt, nach der Geburt ihres Kindes, benutzt sie die Menstruationsbinden, die sie zu Hause hat, als Windeln für das neugeborene Baby.

Mit diesen gesammelten Aussagen zur Nutzung unterschiedlicher Menstruationsprodukte möchten wir verdeutlichen, welche Faktoren die Auswahl beeinflussen – von den praktischen Überlegungen zu den Möglichkeiten, die Produkte wie Tampons eröffnen, der Vertrautheit mit bestimmten Produkten, lokalen Vorstellungen von Jungfräulichkeit bis hin zum Zugang zu und den Kosten der Produkte. In Fluchtkontexten haben die Hilfsorganisationen, die in den Camps aktiv sind, einen großen Einfluss auf die Praktiken der Frauen. Die humanitären Organisationen verteilen Produkte, deren Auswahl häufig auf Vorannahmen zu den lokalen Praktiken basiert (Schmitt et al. 2017: 4). Die Menschen in den Camps sind dann abhängig von dieser Verteilung, die häufig nicht regelmäßig erfolgt oder nach einiger Zeit aussetzt, da davon ausgegangen wird, dass dann wieder auf herkömmliche Praktiken und Ressourcen zurückgegriffen wird. Welche Produkte verwendet werden, wird, ebenso wie außerhalb von Fluchtkontexten, von Trends beeinflusst, die von den Menschen unter an-

plains: "They bring us menstrual pads. That's really good. They are meant to keep infections away. But that's a lost cause, as long as the conditions in the toilets remain a catastrophe."

Marzieh: Because she has not begun menstruating again after giving birth, she uses the menstrual pads she has at home as disposable nappies for the new-born baby.

With these collected statements on the use of different menstrual products, we want to make evident the factors that influence women's choices, which include: practical considerations regarding the possibilities offered by products such as tampons; familiarity with certain products; local attitudes towards and perceptions of virginity; and issues of access to and the cost of these products. In refugee contexts, the aid organisations active in the camps influence the practices of the menstruators greatly. The selection of the products distributed by the humanitarian aid organisations is often based on assumptions about local practices (Schmitt et al. 2017: 4). The people in the camps depend on these distributions, which are often irregular or end after a certain period of time, with distributors surmising that conventional practices and resources can be accessed again. A further major factor impacting the products used, both within and outside of refugee contexts, are the trends people follow – on Facebook, Instagram or other social media and in other societal locations. Thus, we emphasise that menstruating

derem mit Hilfe sozialer Medien, wie Facebook oder Instagram, verfolgt werden. Wir möchten deswegen an dieser Stelle betonen, dass Menstruierende auf der Flucht nicht passiv in der Auswahl der Menstruationsprodukte sind, sondern sich beispielsweise über die Nutzung von Menstruationstassen informieren, und dass sich daher lokale Praktiken ständig verändern (Tellier 2020: 596).

Nachhaltigkeit ist ein weiterer Faktor, der bei der Auswahl der verwendeten Produkte sowohl von Seiten der Nutzer:innen als auch der humanitären Organisationen eine Rolle spielt. Wiederverwendbare Produkte wie Menstruationstassen oder waschbare Einlagen aus Stoff reduzieren den Müll und müssen nur einmal angeschafft werden. Allerdings benötigt die Nutzung dieser Artikel weitere praktische Maßnahmen, da Menstruationstassen regelmäßig abgekocht und wiederverwendbare Binden gewaschen und getrocknet werden müssen (Tellier 2020: 602). Dabei gilt es zu beachten, dass auch mit dem Waschen und Trocknen der Produkte Schamgefühle verbunden werden und es oftmals keine Gelegenheit gibt, die Stoffbinden geschützt von den Blicken anderer zu trocknen. Teilweise werden die Binden dann unter anderen Kleidungstücken versteckt aufgehängt oder unter Matratzen gelegt. Das Trocknen dauert dann wesentlich länger, sodass eine größere Anzahl an Binden gebraucht wird, die häufig nicht vorhanden ist. Wenn die Binden nicht gut gewaschen und getrocknet werden können, kann die Nutzung zu Ausschlägen und Infektionen führen (Schmitt et al. 2017: 5).

Die Nutzung von Einwegprodukten verlangt wiederum geeignete Entsorgungsmöglichkeiten, die nicht immer ausreichend zur

people who are fleeing are not passive in their selection of menstrual products; they inform themselves, for example, about the use of menstrual cups, with the result that local practices change constantly (Tellier 2020: 596).

Sustainability is a further aspect in how products are selected, both from the perspective of users and from humanitarian organisations' perspectives. Reusable products such as cups or washable fabric inserts reduce waste and must be purchased only once. However, the use of these items requires other practical measures; cups require regular boiling and reusable sanitary pads must be washed and dried (Tellier 2020: 602). It should be noted that washing and drying the products can also lead to feelings of shame, as there is often no place to dry sanitary pads out of sight. Sometimes these pads are then hung out to dry hidden under other items of clothing or are placed under mattresses. These then take much longer to dry, meaning that a larger number of sanitary pads are needed, which are often simply not available. If the reusable pads cannot be washed and dried properly, their use can lead to rashes and infections (Schmitt et al. 2017: 5).

The use of single-use products, in turn, requires disposal options – and there are often not enough of these. Moreover, these items must be purchased by the users themselves, after the aid organisations stop distribution. Because most people who are fleeing only have limited finances, alternative coping strategies are developed when menstrual products become too ex-

Verfügung stehen. Außerdem müssen diese Artikel nach dem Aussetzen der Verteilung durch die Hilfsorganisationen selbst gekauft werden. Da den meisten Menschen auf der Flucht nur wenige finanzielle Mittel zur Verfügung stehen, werden andere Bewältigungsstrategien entwickelt, wenn Menstruationsprodukte zu teuer sind. Beispielsweise werden Stoffstreifen aus verfügbaren Kleidungsstücken gerissen, wie es die Frauen in der Geschichte von Selma taten.

Wie wir hier darlegen, ziehen die verwendeten Produkte jeweils eine Reihe praktischer Konsequenzen nach sich, die von den humanitären Organisationen bedacht werden müssen.

Toiletten und andere Orte

Wie Malihehs Gesprächspartnerin Havva erwähnte, sind Produkte wie Binden nur ein Faktor, der beachtet werden muss, wenn wir uns mit dem Thema Menstruation und Flucht/Migration beschäftigen. Ein weiterer wichtiger Aspekt sind die sanitären Anlagen, die zur Verfügung stehen, und deren Sauberkeit. Das folgende Schaubild verdeutlicht die unterschiedlichen Elemente, die eine ›frauenfreundliche Toilette‹ im humanitären Kontext ausmachen. Vieles davon erscheint selbstverständlich: Licht, ein Mülleimer, ein Schloss an der Tür. Aber wie so vieles im Kontext von Migration und Flucht, sind diese Dinge alles andere als selbstverständlich. Sanitäre Anlagen in Lagern sind häufig nicht privat, nicht abschließbar, haben kein Licht, sind nicht für Männer und Frauen getrennt. Es gibt häufig kein Wasser für das Waschen von Binden und Kleidung. Einwegprodukte

pensive. As the women in Selma's story demonstrated, for example, strips of fabric can be torn from old clothing.

As we explain, above, each group of menstruation products generates a range of practical consequences that must also be considered by humanitarian organisations.

Toilets and Other Spaces

As Maliheh's interview partner, Havva, explained, products like sanitary pads are just one factor that needs to be considered when dealing with menstruation during flight and migration. Another important aspect are available the sanitary facilities and their cleanliness. Image 5 illustrates the different elements that make up a 'female-friendly toilet' in a humanitarian aid context. Many of the items listed seem self-evident: lighting, a rubbish bin, a lock on the door. But like so much in the context of migration and displacement, the presence of such elements is anything but self-evident. Sanitary facilities in refugee camps are often not private, cannot be locked, have no lighting and are not separate for men and for women. There is frequently no running water for washing sanitary pads or clothing. Disposable products are sometimes thrown away in the toilets, to stop them being seen in the garbage cans.

Safety is an important aspect of female-friendly toilets, which is not guaranteed in most refugee camps. Image 6 shows the toilets in the Moria camp. When

Abb. 5: Eine frauenfreundliche Toilette aus *A Toolkit for Integrating Menstrual Hygiene Management (MHM) into Humanitarian Response*, 2017, Columbia University und International Rescue Committee. S. 41

Image 5: A female-friendly toilet, taken from *A Toolkit for Integrating Menstrual Hygiene Management (MHM) into Humanitarian Response*, 2017, Columbia University und International Rescue Committee, p. 41

werden teilweise in den Toiletten entsorgt, um zu verhindern, dass sie in den Mülleimern für alle sichtbar sind (Tellier 2020: 596).

Sicherheit ist ein wichtiger Aspekt von frauenfreundlichen Toiletten, was in den meisten Lagern nicht gewährleistet ist. Auf dem Foto sind Toiletten in Moria zu sehen. Wenn die Türschlösser kaputt sind, bringen Frauen ein Kind mit, das dann vor der Tür wartet, um sicherzustellen, dass niemand die Tür öffnet.

Unser Blick auf die unterschiedlichen Materialitäten der Menstruation im Kontext von Flucht und Migration zieht unsere Aufmerksamkeit also auf alltäglich Dinge wie

the locks are broken, women often bring a child with them, who then waits outside the door to make sure no one opens it.

Our discussion of the different materialities of menstruation in the context of flight and migration has drawn attention to everyday issues such as toilets, pads and tampons. It has become clear that access to these materials of everyday life – be it clothing, a secure roof over one's head or even sanitary pads cannot be taken for granted; indeed, flight and migration makes accessing them even more precarious. Feelings of shame and stigma associated with menstruation further complicate

Menstruation Products

Abb. 6: Toiletten im Lager Moria, März 2020.
©Kiki-Alpha

Image 6: Toilets in the Moria Camp, March 2020.
©Kiki-Alpha

Toiletten, Binden und Tampons. Es wurde deutlich, dass der Zugang zu diesen Materialitäten des Alltags – sei es Kleidung, ein sicheres Dach über dem Kopf oder eben Binden – keine Selbstverständlichkeit ist und durch Flucht und Migration erschwert wird. Wie von uns beschrieben, verkomplizieren Schamgefühle und Stigmata, die mit der Menstruation verbunden werden, den Zugang noch weiter, da häufig nicht offen über Bedürfnisse oder Praktiken gesprochen wird.

Die Beschäftigung mit diesen alltäglichen Dingen verdeutlicht außerdem, wie persönliche Erfahrungen während der Flucht oder Migration von Faktoren wie Gender, Alter oder

access to these products, because needs and practices are often not discussed openly.

Moreover, engaging with these everyday things makes concrete how personal experiences during flight and migration are characterized by factors such as gender, ages and motherhood. In doing so, this engagement contributes to a more differentiated perspective on fleeing and migration.

Menstruationsprodukte

Abb. 7 und 8: Dusche und Toilettenräume im Durchgangslager Friedland, Februar 2019. ©Maliheh Bayat Tork

Image 7 and 8: Shower and toilet rooms in the Transit Camp Friedland, February 2019. ©Maliheh Bayat Tork

auch Mutterschaft geprägt werden, und trägt somit zu einem differenzierteren Blick auf Flucht und Migration bei.

Literatur | References

Green-Cole, Ruth. 2020. Painting Blood: Visualizing Menstrual Blood in Art. In: Chris Bobel, Katie Ann Hasson, Elizabeth Arveda Kissling, Tomi-Ann Roberts: *The Palgrave Handbook of Critical Menstruation Studies*, 787–801. Singapur.

Hawkey, Alexandra J.; Jane M. Ussher; Janette Perz; Christine Metusela. 2017. Experiences and Constructions of Menarche and Menstruation among Migrant and Refugee Women. *Qualitative Health Research* 27 (10): 1473–1490.

Hawkey, Alexandra J.; Jane M. Ussher; Janette Perz. 2020. »I Treat My Daughters Not Like My Mother Treated Me«: Migrant and Refugee Women's Constructions and Experiences of Menarche and Menstruation. In: Chris Bobel, Katie Ann Hasson, Elizabeth Arveda Kissling, Tomi-Ann Roberts: *The Palgrave Handbook of Critical Menstruation Studies*, 99–113. Singapur.

Schmitt, Margaret L.; David Clatworthy; Ruwan Ratnayake; Nicole Klaesener-Metzner; Elizabeth Roesch; Erin Wheeler; Marni Sommer. 2017. Understanding the Menstrual Hygiene Management Challenges Facing Displaced Girls and Women: Findings from Qualitative Assessments in Myanmar and Lebanon. *Conflict and Health* 11 (19): 1–11.

Sommer, M., Schmitt, M., Clatworthy, D. 2017. A Toolkit for Integrating Menstrual Hygiene Management (MHM) into Humanitarian Response. (1st ed.). New York: Columbia University, Mailman School of Public Health and International Rescue Committee. https://www.publichealth.columbia.edu/sites/default/files/pdf/mhm-emergencies-full-toolkit.pdf

Tellier, Marianne; Alex Farley; Andisheh Jahangir; Shamira Nakalema; Diana Nalunga; Siri Tellier. 2020. Practice Note: Menstrual Health Management in Humanitarian Settings. In: Chris Bobel, Katie Ann Hasson, Elizabeth Arveda Kissling, Tomi-Ann Roberts: *The Palgrave Handbook of Critical Menstruation Studies*, 593–608. Singapur.

Ussher, Jane M. 2006. *Managing the Monstrous Feminine: Regulating the Reproductive Body*. London.

Anmerkungen | Notes

1 Dieser Begriff weist darauf hin, dass nicht alle Mädchen und Frauen menstruieren (z. B. während der Schwangerschaft und nach der Menopause) und dass nicht alle Menstruierenden Frauen sind (z. B. Transmänner) (Tellier et al. 2020: 594).

2 Diese Information basiert auf der Bereitstellung von Hygienepaketen im Durchgangslager Friedland und in Lagern in Maschhad, Iran.

3 Die Verbindung zwischen Tampons und Jungfräulichkeit entsteht durch Fehlinformation rund um das ›Jungfernhäutchen‹, auch Hymen genannt, die in vielen Kontexten weitverbreitet sind und nicht der biologischen Realität entsprechen. Dennoch beeinflussen sie die Auswahl von Menstruationsprodukten.

1 With this nonspecific terminology we want to do justice to the fact that not all girls and women menstruate (i.e., during pregnancy or after menopause), and that not all menstruating people are women (e.g., transmen) (Tellier et al. 2020: 594)

2 This information is based on the provision of 'hygiene packets' in Friedland Transit Camp in Germany and Camps in Maschhad, Iran.

3 This connects with the idea that these products break the hymen. Misconceptions about the hymen are widespread in many contexts and do not correspond to biological reality. Yet, they influence the choice of menstrual products nonetheless.

Das Smartphone:
Überlebenswerkzeug und mobile Heimat

The Smartphone:
Survival Tool and Mobile Sense of Home

Andrea Lauser | Antonie Fuhse |
Maliheh Bayat Tork | Miriam Kuhnke

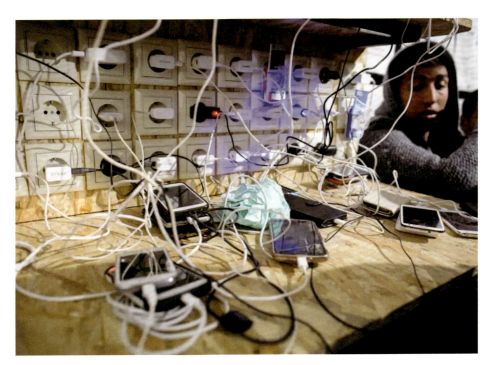

Abb. 1: Smartphones werden im Camp in der Nähe von Idomeni, nahe der Grenze zwischen Mazedonien und Griechenland, aufgeladen, März 2016. ©Kay Nietfeld/epa

Image 1: Smartphones charging in the camp near the village of Idomeni, Greece, not far from the Greek-Macedonian border, March 2016. ©Kay Nietfeld/epa

Das Smartphone ist in der digital-technologischen Wirklichkeit des 21. Jahrhunderts ein steter Gefährte oder geradezu ein Körperteil des Menschen geworden. Das Gerät, das wir

In the digital and technological reality of the 21st century, the smartphone has become a constant companion – or even an additional human body-part, as some

mannigfaltig nutzen, ist gleichzeitig ein Ort, in dem wir leben (Miller et al. 2021). Das flache rechteckige Objekt aus Kunststoff, Silikon sowie seltenen Metallen und Erzen passt bequem in die Tasche einer Jeans. Tatsächlich wurde dieses handliche Ding mit dem Potenzial eines multimedial vernetzten Universalcomputers erst in den frühen 2000ern auf den Markt gebracht. Seitdem komprimiert und transformiert es die Beziehung der Nutzer:innen zu ihrer Umgebung und ist weltweit tief in das Leben von Abermillionen Menschen eingedrungen (Eriksen 2020).

In diesem Sinne sind Smartphones unterwegs, besonders auch auf Fluchtrouten, vieles: Sie sind Überlebenswerkzeuge, die sowohl Möglichkeiten als auch Risiken in sich bergen; sie bilden individuelles Leben ab zwischen distanzierter Intimität und (Selbst-)Repräsentation; sie dienen als Archiv, Bibliothek, Kino, Disco, Spielsalon, Erinnerungsspeicher, Übersetzungshilfe, Reisekompass und mobile Heimat. Das Smartphone als mobile Heimat – »transportal home« (Miller et al. 2021: xi) –, das mag zunächst widersprüchlich klingen, da in der Regel Heimat mit Ortstreue und Immobilität verbunden wird. Doch das Smartphone ist tatsächlich in der Lage, die verschiedenen Komponenten eines Zuhauses zusammenzuführen. Nutzer:innen machen vielfach genau dies und verwenden das Gerät als ›Ort‹ der Stabilität, der Selbstvergewisserung und für die Vergegenwärtigung der eigenen Existenz.

Für Menschen auf der Flucht ist ein Mobiltelefon – beziehungsweise ein Smartphone, also explizit kein einfaches Tastenhandy, sondern heutzutage ein Telefon mit Touchscreen, das an das globale Netz angebunden ist – alles andere als Luxus, wie bisweilen argwöhnisch

would claim. The device, which we use in such multifarious ways, is simultaneously a place we live in (Miller et al. 2021). The flat, rectangular object is made from synthetic materials, precious metals, and various ores and fits comfortably into a jeans' pocket. It's sometimes worth remembering that this handy thing containing the potential of a multimedia, networked, universal computer only first hit the market in the early 2000s. Since then, it has condensed and transformed relationships between its users and their environment, and has dug deep into the lives of countless millions of people worldwide (Eriksen 2020).

In this sense, smartphones in transit, and particularly on routes of flight, are a myriad of things: They are survival tools that harbour both risks and possibilities. They model individual life, in a space between distanced intimacy and (self-)representation. They serve as archives, libraries, cinemas, discos, amusement arcades, memory reservoirs, translation assistants, travel compass and a mobile sense of home. The smartphone as a "transportal home" (Miller et al. 2021: xi) may sound like a contradiction at first, because a sense of home is normally connected to loyalty to a single location and immobility. But the smartphone really is capable of bringing together the various components that constitute a feeling of home. And users frequently exploit this very possibility, using the device as a 'location' of stability, of self-reassurance and of realising their own existence in the present.

For people who are fleeing, a mobile phone – with which is meant a smart-

unterstellt wird. Ein Smartphone mit seinem Globalen Positionsbestimmungssystem (GPS) und vielen verschiedenen Anwendungen (Apps) ist für Menschen auf der Flucht genauso wichtig wie ihre Kleidung oder das Boot, in das sie steigen, um etwa über das Mittelmeer nach Europa zu gelangen. Unterwegs auf Fluchtrouten gelten Smartphone, Nahrung und Wasser als Grundbedürfnisse (siehe auch Brunwasser 2015).

Das Smartphone als ›Lebensader‹ – ebenso wichtig wie Wasser und Nahrung – dient unterwegs der Planung, Navigation und Dokumentation. Es ermöglicht den regelmäßigen Kontakt mit Familie, Freunden und mehr oder weniger informellen Transportunternehmen (oder sogenannten Schmugglern und Schleusern) und/oder denjenigen, die ihnen helfen, die Reise zu planen und zu organisieren. Telefonanrufe und Nachrichten machen aber nur einen kleinen Teil des existenziellen Potenzials eines Smartphones aus. Apps, Websites, soziale Medien, Navigationstools und Übersetzungsdienste, Kamera und Audioaufnahmen bilden zusammen eine digitale Infrastruktur, die zu einem integralen Bestandteil jeder (Flucht-)Reiseroute geworden ist.

Über das Telefon erhalten Menschen auf der Flucht Informationen über Routen und Transportkosten. Sie können erfahren, welche Grenzen offen und welche geschlossen sind, ja das Smartphone kann ihnen das Leben retten, indem es über die Wetterbedingungen vor einer Meeresüberquerung informiert oder die GPS-Daten bei Seenot an helfende Mediator:innen weitergibt (Gillespie 2016b). Mit der Preisgabe der eigenen Geodaten bietet ein Smartphone im doppelten

phone, and explicitly not a push-button mobile, but rather, today, a phone with a touchscreen, connected to the worldwide web – is anything but a luxury, as has often been jealously insinuated. Rather, for people fleeing, a smartphone with its global positioning system (GPS) and lots of different apps is just as important as their clothing or the boat they climb into to go to Europe across the Mediterranean. In transit, on routes of flight, a smartphone, food and potable water are regarded as basic needs (see also Brunwasser 2015).

Functioning as a 'lifeline', and considered just as important as water and nutrition, the smartphone serves the purposes of planning, navigation and documentation for people on the move. It facilitates regular contact with family and friends, with the transport enterprises – some more formal, some less so – run by people who frequently get called '(people) smugglers', and with others, who help those fleeing to plan and organise their journeys. But phone calls and phone messages only constitute a small part of the existential potential of a smartphone. Apps, websites, social media, navigation tools, translation services, photographs and audio-recordings come together to form a digital infrastructure, which has become an integral component of every travel route, whether for journeys of flight or for other purposes.

People who are fleeing receive information via their phones regarding routes and transport costs. They can learn which borders are still open and which already closed; indeed, a smartphone can save their lives, by informing them about weath-

Sinne unbegrenzte Informationen. Damit gehen aber auch Gefahren neuer Formen der Ausbeutung und Überwachung einher, da die »digitale Passage« (Gillespie et al. 2018) von einer Vielfalt sozio-technischer und materieller Zusammenhänge bestimmt wird – beispielsweise, wenn die Routen über informelle Transportunternehmen finanziert und durch ein Grenzregime kriminalisiert werden.

Nachfolgend möchten wir anhand ausgewählter Beispiele diese Dynamik zwischen Grundbedürfnis, Chance und Verwundbarkeit, zwischen Resilienz und Solidarität im Kontext von Flucht und digitaler Verbindung illustrativ erzählen.[1] Als drei wesentliche durch das Smartphone bereitgestellte und miteinander verbundene Dienste betont der Ethnologe Eriksen die der *Lokalisierung*, der *Vernetzung* und der *Mikro-Koordination* (Eriksen 2020). Um die ›Vielsprachigkeit‹ und Vielseitigkeit von Smartphones allgemein als auch besonders unterwegs – und hier speziell auf der Flucht – zu verstehen, illustrieren wir diese genannten Grunddienste über die Stationen a) Digitale Infrastruktur, b) Info-Center und Reiseagentur, c) Kommunikation und Verbundenheit bzw. Nähe und Distanz, die sich an Eriksens Ausführungen anlehnen, gehen jedoch mit dem vierten Aspekt der d) (Selbst-)Dokumentation per Smartphone darüber hinaus.

Digitale Infrastruktur – Mobile Ladegeräte – Stromzugang ist Handlungszugang

So essenziell Smartphones auf der Reise sind, gleichzeitig sind sie keine ›robusten‹, sondern

er conditions before a sea crossing, or by sending their GPS coordinates in a sea emergency to helpful and mediating persons or agencies (Gillespie 2016b). Through divulging an individual's geodata, a smartphone can provide information that crosses national and other forms of borders. However, this is accompanied with dangers manifesting as new forms of exploitation and surveillance, because what has been called the "digital passage" (Gillespie et al. 2018) is determined by a range of socio-technological and material interrelations. This is the case, for example, when the journeys of flight are financed via informal travel enterprises, which are criminalised by a border regime.

In the following, we wish to narrate, in an illustrative manner and using selected examples[1], these dynamics, which encompass basic needs, opportunity and vulnerability, and which stretch to link resilience and solidarity in the context of flight and digital connection. The anthropologist Thomas Eriksen emphasises three essential distinctive yet connected "affordances" smartphones place at their users' disposal: "*location, networking*, and *micro-coordination*" (Eriksen 2020, emphasis in original). To enrich our understanding of the 'multilinguality' and manifoldness of smartphones – in general, particularly while in transit and especially during flight journeys – we illustrate the aforementioned affordances via the following virtual and/or physical stations: a) digital infrastructure; b) information centres and travel agencies; and c) communication and connectedness/proximity and distance.

eher ›fragile‹ und ›pflegebedürftige‹ Reisebegleiter. Ebenso wie die Reisenden sich nach dem Ankommen in den Unterkünften erfrischen und reinigen und sowohl ihre Nahrung als auch ihre Kleidung erneuern, so müssen auch Smartphones aufgefrischt, gereinigt, aufgeladen und repariert werden (Abbildung 1). Es gilt, Strom, WiFi und passende SIM-Karten zu organisieren. Einige humanitäre Unterkünfte und Flüchtlingslager reagieren auf dieses Grundbedürfnis, indem sie entsprechend für (solarbetriebene) Ladegeräte und freien WiFi-Zugang sorgen.

Wer die Grenze zu Fuß überquert, insbesondere diejenigen, die keinen Routenplan haben, setzen alles daran, ihre Smartphones am Leben zu erhalten. So wird auf verschiedene Weise versucht, den Energieverbrauch zu minimieren und für zusätzlichen Strom zu sorgen, beispielsweise durch das Deinstallieren unnötiger Anwendungen auf dem Smartphone, das Aktivieren des Energiesparmodus sowie das Mitnehmen von Ersatz-Akkus und mehrerer Powerbanks. Javad, ein junger Mann aus Afghanistan, der die Grenze von der Türkei nach Griechenland zu Fuß überquert hat, erklärte Maliheh Bayat Tork, Mitarbeiterin im Projekt »Zur Materialität von Flucht und Migration«, in einem Telefonat via Telegram vom 20. März 2021:

> … all das wird deinen [Smartphone-]Akku dennoch nur für drei Tage schonen. Wenn kein WiFi für die Navigation zur Verfügung steht, musst du dein Smartphone auf ›mobile Daten‹ einschalten und das frisst deinen Akku sehr schnell auf. Ich hatte die Google-Karte im Offline-Modus im Voraus herunterge-

These three stations build on Eriksen's deliberations but are examined together with a fourth aspect: d) (self)-documentation using smartphones, which reaches beyond the terrain covered by Eriksen.

Digital Infrastructure – Mobile Chargers – Accessing Electricity Means Accessing Agency

As essential as smartphones are while travelling, they are not robust travel companions but rather fragile ones, often in need of care. Just as travellers freshen up, wash and renew their food as well as their clothing supplies on arriving at their accommodation, smartphones too must be freshened up, cleansed, charged and repaired (Image 1). The imperative is to organise electricity, wifi and SIM cards appropriate to the current location. A large number of humanitarian forms of accommodation and refugee camps react to this basic need, by providing chargers, some of which are solar powered, and free access to wifi.

Whoever crosses an international border on foot, and especially those with no route plan in print form, will do everything to keep their smartphones 'alive'. Refugees employ a range of means to minimise energy use and provide for additional electricity, including deinstalling unnecessary apps, activating the battery-saving mode and carrying a number of replacement batteries and power banks with them. But as Javad, a young, Afghan man who crossed the border between Turkey and Greece on foot, explained to Maliheh Bayat

laden; das weiß nicht jeder, aber es gibt dennoch ein Problem, wissen Sie, wenn die Offline-Karte nicht aktualisiert ist, dann kann sie ungenau sein. Es ist also riskant.

Reparatur und ›Reinigung‹

Im Vergleich zu Kleidung und Lebensmittel kann das Smartphone sogar an allererster Stelle der Grundbedürfnisse rangieren. Das zeigt sich besonders dann, wenn das Reisegepäck drastisch eingeschränkt werden muss, wie zum Beispiel bei einer überfüllten Bootspassage. Kleidung und Lebensmittel lassen sich bei der Ankunft relativ günstig ersetzen. Nicht so ein Smartphone mit all seinen gespeicherten Informationen und Kontakten. Zu wertvoll ist das Smartphone in seiner funktionalen und emotionalen Vielseitigkeit, um es freiwillig aufzugeben. Und wenn es Schaden genommen hat – durch Wasser oder Bruchstellen – ist ein Smartphone-Reparateur ein Segen.

Im Lager Moria in Griechenland nahm im Juni 2020 ein Mann die Rolle des Reparateurs ein, wie Maliheh über einen Bekannten erfuhr,

Tork – staff member in the project "On the Materiality of (Forced) Migration" – in a Telegram phone call on 20 March, 2021:

"... all that will only protect your [smartphone] battery for three days. If you have no wifi at your disposal for navigation, you're forced to turn on your phone's mobile data, and that eats away your battery very quickly. I had downloaded the Google Map into offline mode in advance; not everyone remembers to do this, but choosing to do so also creates problems, you know – if the offline map hasn't been updated, it can be imprecise. So, it's a risky [strategy] too".

Repair and 'Cleansing'

On a level comparable with food and clothing, the smartphone can, on occasion, rise to first place in the things humans require to meet their basic needs. This becomes especially evident when travel baggage has to be reduced drastically, before or during a crossing in an overcrowded boat, for example. Clothes and food can be replaced relative-

Abb. 2: Ein Mann (Name nicht bekannt) im Lager Moria repariert ein Smartphone, Juni 2020. Foto von Ali Shams Eddin, einem syrischen humanitären Freiwilligen, der in der »Movement On The Ground« organisation in Moria mitarbeitet. ©Ali Shams Eddin

Image 2: A man, name unknown, repairs a smartphone in Moria Camp in June 2020. Photo by Ali Shams Eddin, a Syrian humanitarian volunteer, who works for the "Movement on the Ground" organisation in Moria. ©Ali Shams Eddin

Abb. 3: Maamun repariert Telefone im Geflüchteten-Camp Zaatari, Jordanien, Juni 2015. ©Pablo Tosco/Oxfam Intermón

Image 3: Maamun repairs telephones in Zaatari Refugee Camp, Jordan, June 2015. ©Pablo Tosco/Oxfam Intermón

der dort in der humanitären Hilfe tätig ist. Täglich reparierte er zu dieser Zeit die Smartphones seiner Mitbewohner:innen, ohne eine finanzielle Gegenleistung zu verlangen (Abbildung 2). Unter den schwierigen Umständen des Lagerlebens war es ihm ein Bedürfnis und Ausdruck seiner Solidarität, seine ›Camp-Community‹, seine Mitbewohner:innen im Lager, mit seinen Fähigkeiten zu unterstützen und mit dem zu helfen, was er kann.

Der Dokumentarfilm *District Zero – What's Hidden Inside the Smartphone of a Refugee?*[2] von 2015 begleitet den syrischen Geflüchteten Maamun Al-Wadi bei seiner Arbeit. Er betreibt in Zaatari in Jordanien, einem der weltweit größten Flüchtlingslager, in einem kleinen weißen Container einen Laden, in dem er Smartphones repariert, Akkus auflädt, Tonaufnahmen, Dokumente, Bilder und verlorene Inhalte wiederherstellt (Abbildung 3). Die Speicherkarten seiner Mitbewohner:innen

ly cheaply on arrival. Not so a smartphone, with all the vital information and contacts saved on it. The Smartphone's functional and emotional versatility makes it too valuable. And if it has been damaged – by water or by fractures to its screen or casing – finding a smartphone technician feels like a blessing.

In the Moria Camp in Greece in June 2020, one man took on the role of smartphone repairer. This was how an acquaintance of Maliheh's, who works in humanitarian aid in the camp, relayed the story. In this period, the man in question repaired the smartphones of his fellow camp-dwellers, without demanding any form of financial compensation for his work (Image 2). Faced with the difficult circumstances of camp life, this man felt the need to support his 'camp community', his fellow inhabitants in Moria, with the abilities he had, thus expressing his solidarity with then

enthalten ihre Vergangenheit in Syrien: Erinnerungen an glückliche und weniger glückliche Routinen, Familienleben, an Krieg, Zerstörung, Flucht. Als besonderen Service bietet er das Ausdrucken der Fotos an. Die Menschen im Camp lassen Fotos von Angehörigen, von ihren Liebsten ausdrucken und nicht die vom Krieg. Der Film zeigt die Bedeutung dieser Fotos für die Menschen, für ihr Selbstverständnis, für ihre Erinnerung an ihr Leben in Syrien.

Signal – Funknetz und WiFi-Verbindungen

Signal heißt das World Press Photo 2014 (Abbildung 4) des amerikanischen Fotografen John Stanmeyer, das dieser für die Zeitschrift *National Geographic* aufgenommen hat. Am nächtlichen Ufer der Stadt Dschibuti als üblichem Zwischenstopp für Migrant:innen auf der Durchreise aus Ländern wie Somalia, Äthiopien und Eritrea halten diese ihre Smartphones in die Höhe, um aus dem benachbarten Somalia eine günstigere Verbindung und ein besseres Signal zu empfangen. Ein Mitglied der World Press Photo Jury beschrieb das Foto als »subtil und poetisch« und betonte, dass es »Themen, die in der heutigen Welt von großer Bedeutung und Besorgnis sind, vermittelt«.[3] Der Fotograf selbst erklärte: »Es verbindet uns alle, es sind nur Menschen, die versuchen, geliebte Menschen anzurufen.«[4] (BBC News Africa 2014) Das Bild evoziert dabei zwei Assoziationen gleichzeitig: Zum einen spielt es auf das prekäre Leben afrikanischer Migrant:innen an und verweist gleichzeitig auf ihren Zugang zu Technologien wie Smartphones.[5]

and helping them in the way he was able to do.

The 2015 documentary titled *District Zero – What's Hidden Inside the Smartphone of a Refugee?*[2] accompanies the Syrian refugee Maamun Al-Wadi while he works. In Zaatari, Jordan, one of the biggest refugee camps in the world, he runs a shop from inside a small, white container in which he repairs smartphones, charges batteries and restores audio recordings, documents, photos and other contents that have been lost or misplaced (Image 3). The memory cards of the camp's inhabitants contain their past in Syria: memories of happy, or less happy, routines, of family life, of war, destruction and flight. One of his special services he offers is photo printing. The camp-dwellers generally get Maamun to print out photos of relatives and loved-ones, and not of war. The film shows what these photos signify to the people who have them printed, for their self-understanding, and for their memories of life in Syria.

Signal – Radio and Wifi Connections

"Signal" was the name American photographer John Stanmeyer chose for the image elected as World Press Photo 2014, which he took for the *National Geographic* (Image 4). On a beach at night in the city of Djibouti, a typical short stop for migrants passing through from countries including Somalia, Ethiopia, and Eritrea, some of these same migrants hold their smartphones high in the air to get lower charges

Javad, der Gesprächspartner von Maliheh Bayat Tork, schildert, wie unerlässlich ein WiFi-Zugang unterwegs ist:

[Nach einem] Jahr in der Türkei, habe ich noch [immer] meine iranische SIM-Karte benutzt. [Denn für die Registrierung einer lokalen SIM-Karte braucht man Ausweispapiere]. Um mit meiner Familie im Iran zu kommunizieren, habe ich mich mit den kostenlosen WiFi-Netzwerken in der ganzen Stadt verbunden. Diese Verbindungen und Kommunikationsmöglichkeiten sind herzerwärmend und überlebenswichtig. Sobald man jedoch

and a better signal from the neighbouring country of Somalia. A juror for the World Press Photo award described the photo as "subtly done, so poetic", and stressed that it conveyed "issues of great gravity and concern in the world today".[3] The photographer himself declared: "It connects to all of us ... It's just people trying to call loved ones" (BBC News Africa 2014). The image evokes two associations simultaneously: It references the precariousness of the lives lived by many African migrants, while also pointing out their access to technologies including smartphones.[4]

Javad, Maliheh Bayat Tork's interview

Abb. 4: Signal, World Press Photo 2014: Am Strand von Dschibuti-Stadt versuchen Migrant:innen, ein günstiges Telefonsignal aus dem benachbarten Somalia zu bekommen, 26. Februar 2013. ©John Stanmeyer

Image 4: "Signal", World Press Photo, 2014: African migrants crowd the night shore of Djibouti city, trying to capture inexpensive cell signals from neighboring Somalia – a tenuous link to relatives abroad. Photo taken 26 February 2013. ©John Stanmeyer

Smartphone

die Grenze überquert hat und in den Bergen und Wäldern unterwegs ist, gibt es keine WiFi-Signale mehr, und aufgrund des Roaming-Service ist die Nutzung mobiler Daten sehr kostspielig. Dies war ein großes Problem, als ich die Grenze von der Türkei nach Griechenland überquerte. Ein WiFi-Signal ist alles, was du möchtest. Wenn man in einer Stadt ankommt, dann sucht man die Nähe eines Restaurants oder Supermarkts, um sich wieder zu verbinden, wie schwach auch immer das WiFi-Signal ist. Das ist das A und O.

›Infocenter‹ und ›Reiseagentur‹

Reiseanleitungen, Wegbeschreibungen, Karten und Tipps, was zu tun ist, um vielversprechende Zielorte zu erreichen, und was weiterhin zu tun ist, wenn diese erreicht sind, werden sowohl über Informationsseiten, interaktive Facebook-, WhatsApp-, Telegram- oder sonstige Social-Media-Gruppen vermittelt als auch ganz besonders über SMS und Nachrichten von Freund:innen und Familienmitgliedern, die die Reise bereits (erfolgreich) gemacht haben. Dieses Vorgehen mag auf den ersten Blick vergleichbar sein mit den unzähligen Reisetipps, die im Prinzip jede Internetrecherche zu Reisezielen generiert. Auf den zweiten Blick ist ein digitales Reiseportal – wie z. B. das fünfsprachige Portal *InfoMigrants*[6] – für irreguläre Migration und Flucht ein achtenswertes, aber nicht immer einfaches Projekt. So beschreibt das *InfoMigrants*-Portal die Schwierigkeiten, das breite Spektrum an verfügbaren Apps zu untersuchen, zu aktualisieren und zu

partner, describes how indispensable wifi access is when people migrate:

"[After a] year in Turkey, I was still using my Iranian SIM card: [identification papers would have been needed to register a local SIM card]. To communicate with my family in Iran, I connected to the free wifi networks throughout the city. These connections and communication possibilities warm your heart and are important simply for survival. But as soon as one has crossed the border and is moving across mountains and through forests, there are no more wifi signals, and accessing mobile data via roaming services is very pricey. That was a big problem after I had crossed the border from Turkey to Greece. [At times like these] a wifi signal is everything you could possibly wish for. When you arrive in a [new] town or city, you look to be near to a restaurant or a supermarket, so that you can connect, no matter how weak the wifi signal is. That's the bottom line on which all else depends."

'Information Centres' and 'Travel Agencies'

Travel instructions, route descriptions, maps, tips about how to reach promising destinations and what still needs to be done when these are reached are communicated via informative webpages, interactive groups on Facebook, WhatsApp, Telegram and other social media groups, and espe-

bewerten. Welche Apps – von Spenden-Apps über Sprachlern- und Übersetzungssoftware bis hin zu privaten Übernachtungsplattformen – sind wirklich hilfreich? Und welche gar gefährlich, da die Identität eines Flüchtenden dadurch gefährdet werden kann, wenn das Smartphone in falsche Hände gerät? Es gibt unzählige Berichte, wonach Smartphones von nicht wohlgesonnenen Grenzregimebehörden beschlagnahmt (gar zerstört) und Bewegungen, Inhalte, Browserverlauf, Nachrichten usw. nachverfolgt werden.[7]

Karten und Lokalisierung

Mit Hilfe des Smartphones können gedruckte Landkarten fotografiert und somit mitgenommen werden (Abbildung 5). Karten-Apps wie beispielsweise Google Maps positionieren die Nutzer:innen visuell nachvollziehbar. Mit den Funktionen von Google Maps lässt sich genau ermitteln, wie Zieldestinationen mit verschiedenen Fortbewegungsmitteln wie Bus, Fähre, Zug, Auto oder zu Fuß erreicht werden können. Bei einer herkömmlichen Karte wiederum kann die Lokalisierung des eigenen Standortes eine Herausforderung sein. Die Genauigkeit der GPS-Standortbestimmung eines Smartphones ist allerdings auch ein zweischneidiges Schwert, das einerseits dazu beiträgt, sich selbst und andere zu orten, und es andererseits durch die Notwendigkeit, online zu sein, Außenstehenden ermöglicht, die Positionsdaten abzufangen. Laut Gillespie und ihrem Forschungsteam (Gillespie et al. 2016) nutzten ein Großteil der interviewten Geflüchteten routinemäßig Google Maps im eingeschalteten Modus, obwohl sie damit gleichzei-

cially via SMS and messages from friends and family members who have travelled on the same route already, sometimes successfully. At first sight, this way of proceeding may seem comparable with the countless travel tips generated, at least in principle, during every internet search for travel destinations. But look again, and it becomes obvious that a digital travel platform – the "InfoMigrants"[5] platform for example, online in five languages – for irregular migration and people fleeing is an admirable project, but not always a straightforward one. The InfoMigrants platform witnesses the difficulties of looking into the wide spectrum of apps available – and of keeping them up to date and evaluating them. Which apps – and the range extends from donation apps, via language-learning and translation software, to private nightly accommodation platforms – really help? Which are actually dangerous, because the identity of a refugee can be threatened if the smartphone falls into the wrong hands? Countless reports relay smartphones being confiscated, and even destroyed, by border regime authorities not well-disposed to migrants. There are also countless reports that the same border regime authorities have spied on physical movements, general digital content, specific web-browser searches and telephone messages.[6]

Maps and Localisation

Smartphones can be used to photograph printed maps, making these more easily portable (Image 5). Map apps, including, for

Abb. 5: Geflüchtete fotografieren eine Karte auf der griechischen Insel Lesbos, 21. September 2015. ©Iakavos Hartziavrou

Image 5: Refugees taking a photo of a map on the Greek island of Lesbos, 21 September 2015. ©Iakavos Hartziavrou

tig auch Gefahr liefen, von staatlichen Stellen lokalisiert und erfasst zu werden (Gillespie 2016a: 15f.; für den ausführlichen Bericht siehe Gillespie et al. 2016). Andererseits kann das Versenden der genauen GPS-Daten in höchster Not zur rechtzeitigen Rettung führen.

Javad beschreibt die Rolle des Smartphones unterwegs als ein Wunder:

> ... auf der gesamten Route haben wir uns mit Hilfe von GPS auf unseren Smartphones orientiert. Die Online-Google-Karte hat zwar viel Akku verbraucht, aber da sie aktualisiert war, hat sie die Polizeistationen genau lokalisiert; so konnten wir diese leicht umgehen. ... Vor Jahren, als noch nicht jeder ein Smartphone hatte, musste man sich beim Grenzübertritt

example, Google Maps, position the users on the map in a way that is visually comprehensible. Such functions can be used to discover precisely how target destinations can be reached using various means of transit: bus, ferry, train, car or even on foot. On a traditional map, by contrast, it can be a challenge for a user to locate where they currently are. The precision with which GPS can locate an individual smartphone is, however, a double-edged sword, a technology that contributes, on the one hand, to being able to locate oneself and others, while, on the other hand, through the necessity of being online, enables external individuals or organisations to capture position data. According to Gillespie and her research team (Gillespie et al. 2016), the

auf ›Schleuser‹ verlassen. Ich erinnere, dass mein Bruder mit Hilfe von solchen ›Schleppern‹ die Grenzen nach Deutschland überquerte. Aber ich habe es mit Hilfe meines Smartphones und Google Maps geschafft. ... Auch für die Überfahrt auf dem Meer sind Smartphones lebensnotwendig. In der Nacht, in der wir das Boot aus Griechenland besteigen sollten, benutzten die Schleuser ihre Smartphones, um die Höhe und Stärke der Meereswellen zu messen. ... Das Leben der Menschen hängt dann tatsächlich von Internet und Smartphones ab. (Javad im Telefonat via Telegram mit Maliheh, Ende März 2021)

In dem kurzen Video »Your Phone Is Now a Refugee's Phone«[8] werden wir aufgefordert, dieses auf unserem eigenen Smartphone zu betrachten. Für etwa drei Minuten soll sich unser eigener Bildschirm zu einem Smartphone eines Menschen auf der Flucht verwandeln. Wir Betrachter:innen sollen für einen kurzen Moment eine gefährliche Überfahrt mit einem Boot unmittelbar nachvollziehen. Textnachrichten und Telefonnachrichten von Familienmitgliedern und auch von unbekannten Personen mit unklaren Motiven erreichen uns. Aus einer Ich-Perspektive werden wir mit instabilen Internetverbindungen und einem sich leerenden Akku konfrontiert. Plötzlich kontaktiert uns jemand über WhatsApp, warnt uns und fordert uns auf umzukehren (Abbildung 6). Wir fragen uns, woher die Person das weiß und ob wir ihr vertrauen können. Wir schreiben, dass ein Kampf im Lager ausgebrochen ist und Risse erscheinen auf dem Display. Am Ende des Videos erscheint ein Fenster auf dem Display und informiert uns: »Die Flüchtlings-

greater part of the refugees interviewed routinely use Google Maps with their location function switched on, although this increases the danger that state authorities will localise their position and thus seize them (Gillespie 2016a: 15 f.; for the comprehensive report see Gillespie et al. 2016). On the other hand, sending exact GPS data in an extreme emergency can save individuals and groups in time.

Javad describes the role of the smartphone while in transit as a miracle:

> ... we oriented ourselves with help from GPS on our smartphones for the whole route. While the online Google map did use up a lot of our batteries, because it was always up to date, it pinpointed the police stations precisely, making it easy to avoid them. ... Years ago, before everyone had a smartphone, people were forced to rely on 'smugglers' after crossing the border. I remember that my brother crossed the border into Germany with the help of such 'smugglers'. But I managed it with the help of my smartphone and with Google Maps. ... Smartphones are also essential to survival during the sea crossings. During the night in which we were meant to climb into the boat leaving Greece, the smugglers used their smartphones to measure the height and the strength of the sea waves. ... Indeed, people's lives really do depend on the internet and smartphones. (Javad in a Telegram phone call with Maliheh, in late March 2021)

Smartphone

Abb. 6: Screenshot aus dem Video »Your Phone Is Now a Refugee's Phone«, 2018. ©BBC Media Action

Image 6: Screenshot from the video "Your Phone Is Now a Refugee's Phone", 2018. ©BBC Media Action

In the short video "Your Phone Is Now a Refugee's Phone"[7], we are challenged to watch this brief film on our own smartphone. Our own screen is meant to transform itself for around three minutes into the smartphone of a person who is fleeing. A dangerous sea crossing ought to become immediately relatable to us as observers, at least for a brief moment. Text and telephone messages from family members, alongside communications from unknown persons with unclear motives. We are confronted, from a first-person-perspective, with unstable internet connections and a battery that is quickly draining. Suddenly someone contacts us via WhatsApp, warning us and telling us to turn back (Image 6). We ask ourselves how the person knows where we are and whether we can trust them. We appear to type into our phone that a fight has broken out in our camp, and cracks appear in our display. At the end of the video a window appears on the display and informs us: "The refugee crisis is not going away." Although the "OK" under this information is pressed, it's impossible to close the window.[8]

krise wird nicht verschwinden.« Das »OK« unter dieser Information wird gedrückt, dass Fenster lässt sich aber nicht schließen.[9]

»Passenger to Turkey« – Mitfahrende:r in die Türkei

Eine unbekannte Handynummer ist das, womit Flüchtende unter Umständen Kontakt aufnehmen müssen. Wenn sie diese anrufen,

"Passenger to Turkey?"

Under specific circumstances, refugees have to contact someone with a mobile number unknown to them. When they

Abb. 7: Öffentliche Werbung von Schlepper:innen. Foto von einem jungen afghanischen Geflüchteten in Mashhad, Iran, 14. Februar 2020. ©Ebrahim Mobarez

Image 7: Public advertising, written by people smugglers. Photo by Ebrahim Mobarez, a young Afghan refugee in Mashhad, Iran, 14 February 2020. ©Ebrahim Mobarez

dann melden sich mehr oder weniger legale Organisationshelfer:innen, mit denen eine unsichere und risikoreiche Reiseroute zu vereinbaren ist.

Auf dem Weg durch eine (Grenz-)Stadt können die Flüchtenden hier und da auf öffentlich an Wänden annoncierte Werbung von ›Reisebüros‹ stoßen. Besonders in Gebieten, die von flüchtenden Migrant:innen bewohnt werden, dienen öffentliche Wände als informelle Werbung mit Angaben von Handynummern zu potenziellen ›Schlepper:innen‹ für den Grenzübertritt.

Dieses Foto (Abbildung 7) wurde in Golshahr, einer der am Stadtrand gelegenen Gegenden Maschhads im Iran aufgenommen, die vor allem von afghanischen Geflüchteten bewohnt wird. Auf der Wand stehen eine Telefonnummer und ein kurzer Text auf Farsi: Die Telefonnummer ist eine türkische Handynummer, darunter steht der Name des Inhabers der Nummer und das Angebot zu einer »Mitfahrgelegenheit die Türkei«. In dieser Gegend ist es nicht schwer, die Nummer von ›Schlepper:innen‹ zu erfahren.

phone such a number, the call is answered by 'helpers' working for various organisations – some more legal, some less so – with whom they need to agree on a route that is unsafe and fraught with potential risks.

When passing through towns, particularly near borders, refugees occasionally see public advertising for "travel agencies" posted onto walls. Particularly in regions inhabited by migrants in flight, public walls serve as informal locations for adverts, with details of mobile numbers of potential 'smugglers' for border crossings.

Image 7 was taken in Golshahr, one of the districts on the edge of Mashhad in Iran which is inhabited primarily by Afghan refugees. A telephone number is written on the wall, accompanied by a short text in Farsi. The number is a Turkish mobile number, under which the name of the number's owner is written, and the offer of a "lift to Turkey". This is a region in which numbers of 'people smugglers' are not hard to come by.

Smartphone

Abb. 8: Geflüchtete am Budapester Keleti-Bahnhof checken ihre Smartphones, 2. September 2015. ©Artur Widak

Image 8: Refugees in Budapest Keleti Railway Station, checking their smartphones, 2 September 2015. ©Artur Widak

Kommunikation und Verbundenheit – Nähe und Distanz

Communication and Connectedness – Proximity and Distance

Ein Smartphone beschleunigt die zwischenmenschliche Vernetzung und ermöglicht einen neuen Rhythmus und eine neue Intensität in der sozialen Interaktion. Über ein Smartphone verschiebt sich das Verhältnis von Nähe und Distanz ebenso wie das, was wir gewohnt sind als ›öffentlich‹ und ›privat‹ zu beschreiben (Miller 2021: 867). So können wir inmitten einer Menschenansammlung in enger physischer Nähe von Personen sein, die sich gleichzeitig mittels Smartphone in ihre eigenen Welten entfernt haben, um dort ver-

A smartphone accelerates interpersonal networks and facilitates a new rhythm and a new intensity in social interactions. A smartphone dislocates the relationship between proximity and distance, and between those quantities that we are used to describing as 'public' and 'private' (Miller 2021: 867). This makes it possible for us, when people gather physically, to be in close physical proximity with people who, simultaneously, have distanced themselves into their own worlds by means of

trauten Beschäftigungen nachzugehen und/oder sich mit Freund:innen und Verwandten über Text- und Bildmedien zu unterhalten.

Das Foto (Abbildung 8) – eines von vielen, die im Budapester Keleti-Bahnhof im Sommer 2015 aufgenommen und weltweit in Zeitungen reproduziert wurden – zeigt junge Männer, die eng aneinandergedrängt hinter einer Balustrade auf behelfsmäßigen Matten liegen, sich im flackernden, bläulichen Licht der Bildschirme ihrer Smartphones ausruhen und die Zeitspanne des Wartens mit Netzwerken, Spielen und Medienkonsum nutzen.

Menschen auf der Flucht sind einer spezifischen Zeitlichkeit ausgesetzt. Sie leben in einem Übergangszustand, in einem legalen Schwebezustand, der in der ethnologischen Literatur oft als eine »Betwixt-and-between«-Gegenwart beschrieben wird, die sich grundlegend von der erhofften Zukunft unterscheidet und besonders durch (mehr oder weniger aktives) Warten charakterisiert ist (Lauser et al. 2022; Eriksen 2020; Turner 2015; Hage 2009): Im (Übergangs-)Lager warten sie auf die Bearbeitung ihres Asylantrags. Sie warten darauf, dass über ihren Antrag auf eine Arbeits- und Aufenthaltserlaubnis in einem neuen Land entschieden wird. Sie warten darauf, nach Europa zu gelangen. Auch wenn das Smartphone die trägen Strukturen von Bürokratien nicht verändert, so bietet es doch Mittel, mit ihnen umzugehen. Die Fähigkeiten des Smartphones in den Händen von Migrant:innen bergen das Potenzial, Beschränkungen des Wartens in Handlungsoptionen umzuleiten. Das Smartphone bietet eine Alternative zur öden Zeit des Wartens. In einer Art ›Smartphone-Zeitlichkeit des Wartens‹ (zwischen virtueller Mobilität und ana-

the smartphone, to pursue familiar pastimes in these worlds, and/or to chat with friends and relatives via text and image media.

The photo shown as Image 8, one of many taken in Budapest Keleti station in the summer of 2015 and reproduced in newspapers worldwide, shows young men lying on makeshift mats, crowded closely beside each other behind railings. They are resting in the flickering, blue light of their smartphones and are whiling away the waiting period by networking, playing and consuming media.

People who are fleeing are exposed to a specific form of temporality. They live in a condition of transition, in a legal limbo that has often been described as a "betwixt-and-between" present in anthropological literature, a form of time fundamentally different to the type of future hoped for and one that is characterised by waiting, in its more – or less – active modes (Lauser et al. 2022, Eriksen 2020, Turner 2015, Hage 2009). Refugees wait in transit camps and other emergency accommodation for their asylum applications to be processed. They also wait for their application for work permits and residency permits in a new country to be decided on. They wait for ways to get to Europe. Even if the smartphone does not, in itself, change the lethargic structures of bureaucracy, it does offer a means of getting around them. The capabilities of the smartphone in the hands of migrants contain the potential to redirect the limitations of waiting into possible courses of action. The smartphone offers an alter-

Abb. 9: Bild der Zeichnung von Banafshe, einem jungen afghanischen Mädchen, aufgenommen in Griechenland. Sie lebt jetzt in Deutschland, April 2020. ©Banafshe Heidari

Image 9: Photo of the drawing by Banafshe, a young Afghan woman, taken in Greece in April 2020. She now lives in Germany. ©Banafshe Heidari

loger Immobilität) werden soziale Netzwerke mobilisiert, Apps genutzt, die Weiterreise organisiert und Pläne geschmiedet – neben vielen anderen Dingen. Allerdings können die ständige Kommunikation und die virtuelle Verbundenheit auch Gefühle von Einsamkeit und Isolation verstärken. Digitale Nähe kann die physische Nähe nicht restlos ersetzen, und die ständige Konfrontation mit der Situation im Herkunftskontext führt zu Gefühlen von Machtlosigkeit (Awad und Tossell 2021: 617–618).

Abbildung 9 ist eine Zeichnung von Banafshe, einer 19-jährigen Afghanin aus dem Lager Moria, die im Jahr 2020 angefertigt wurde. Als ihr Vater von den Taliban getötet wurde, floh Banafshe mit ihrer Familie aus Afghanistan in den Iran, wo sie ein Mathematikstudium absolvieren konnte. 2019 verließ sie den Iran mit ihrer Mutter und ihrem älteren Bruder. Seitdem lebten sie in Moria. Ein Feuer, das das Flüchtlingslager Moria im September 2020 fast vollständig verwüstete, haben die drei überlebt. Aber alle Zeichnungen und die Mal-Utensilien von Banafshe wurden von den Flammen verbrannt. Das Bild existiert

native to the dreary waiting periods. In a kind of 'smartphone temporality of waiting', located somewhere between virtual mobility and analogue immobility, social networks are mobilised, apps used, the onward journey organised and plans forged – alongside many other activities. On the other hand, constant communication and virtual connectivity can also accentuate feelings of loneliness and isolation. Digital proximity cannot replace physical closeness indefinitely, and refugees constantly being confronted with the situations in their places and countries of origin experience feelings of powerlessness (Awad and Tossell 2021: 617–168).

Image 9 shows a drawing undertaken in 2020 by Banafshe, a 19-year-old Afghan woman, in the Moria Camp. After the Taliban killed her father, Banafshe fled with her family from Afghanistan to Iran, where she was able to complete a university degree in maths. She left Iran with her mother and with her elder brother in 2019, and after that they lived together in Moria. All three of them survived the fire in September 2020 that almost completely destroyed the camp. But all of Banafshe's drawings, and her drawing and painting utensils, were devoured by the flames. The picture now only exists in her smartphone's

jetzt nur noch in dem Fotoarchiv ihres Smartphones. Das Bild hat Banafshe mit folgenden Worten kommentiert:

> Nun, es hat verschiedene Bedeutungen. Wenn ich weit weg von den geliebten Menschen bin, ist das Mobiltelefon der einzige Ansprechpartner und Zufluchtsort, um in Verbindung und Kontakt zu bleiben. Für mich zeigt diese Zeichnung Heimweh und Einsamkeit. Aber seit wir zu Flüchtlingen geworden sind, und besonders seit ich in Moria weile, habe ich dort viele Menschen aus anderen Ländern kennengelernt. Unter anderen Umständen hätten wir uns vermutlich nie kennengelernt. Und nun helfen wir uns gegenseitig – ohne Erwartungshaltung, sondern weil wir uns gegenseitig unterstützen wollen. Nun bedeutet die Zeichnung für mich also auch, Menschen auf der ganzen Welt die Hand zu reichen und mit ihnen zu kommunizieren, unabhängig von Grenzen und rassistischer Diskriminierung. (Banafshe im Chat via Instagram mit Maliheh am 15. September 2020)

photo archive. Banafshe commented on the picture as follows:

> Well, the photo means many different things. When I'm far away from my loved-ones, the mobile phone [feels like the] only person I can talk to, like a place of refuge, in order to stay connected and in touch. For me, this drawing shows homesickness and loneliness. But since we became refugees, and particularly since I've lived in Moria, I have gotten to know lots of people from different countries. I would presumably never have gotten to know them under other circumstances. And now we help each other – not with an attitude of expectation, but rather because we want to give each other mutual support. So now the drawing also means for me to stretch out our hands to people all over the world, and to communicate with them, regardless of borders and of racist discrimination. (Banafshe during an Instagram chat with Maliheh on 15 September, 2020)

Dokumentation per Smartphone – Erinnerungsspeicher

Reza beschreibt in auf Farsi den Moment im Jahre 2015, als er das Foto auf dieser Seite (Abbildung 10) gemacht hat:

> Ich dachte, es ist vielleicht das letzte Mal, dass ich Afghanistan (Herat) sehen wer-

Documenting Things Using Smartphones – Reservoirs of Memories

Talking in Farsi, Reza described the moment in 2015, in which he took the photo printed in Image 10.

> "I thought that this might be the last time that I ever see Herat in Afghan-

Abb. 10: Selbstportrait von Reza Heidari, einem afghanischen Geflüchteten im Iran, 7. Oktober 2020. ©Reza Heidari

Image 10: Self-portrait by Reza Heidari, an Afghan refugee in Iran, 7 October 2020. ©Reza Heidari

de. Also wollte ich so viele Eindrücke wie möglich festhalten. Ich hatte nur mein Smartphone zum Fotografieren dabei. Und dieses Foto spiegelt eher ein Gefühl wider, das ich in diesem Moment hatte, als eine gezielte Aufnahme. Ich fand das Porträt des Mannes an der Wand, der von den Spiegeln zurückgehalten wird, interessant.

Und humorvoll fährt er fort: »Mit einer Prise Narzissmus versehen, fand ich Gefallen an der Verbundenheit, die die Reflexion meines Gesichts in den beiden Spiegeln mit dem Porträt erzeugte, also nahm ich es als Erinnerungsstück mit.« Er scherzt: »Wir nehmen sogar unsere Erinnerungsfotos ernst.« (Reza im Telefonat mit Maliheh, März 2021)

Menschen verwandeln ihre Smartphones in ein Gerät entsprechend ihrer Persönlichkeit und ihrem eigenen Selbstverständnis. Es ist diese individuelle Bearbeitung, wie z.B. die Speicherung von Daten, Briefen, Mails, Messages, Fotos und Videos sowie das Downloaden von Apps, Musik usw., die das Smartphone in die Lage versetzt, eine erstaunliche

istan. So, I wanted to record as many impressions as possible. The only thing I had with which to photograph was my smartphone. And this particular photo reflects more a feeling that I had in that moment than a purposeful recording. I was interested in the portrait of the man on the wall, who is held back by the mirrors."

And he continued to reflect on the photo in a humorous way: "With a pinch of narcissism in the mix, I was pleased with the connection, which the reflexion of my face in both mirrors built up with the portrait, so I took that as a memento with me …" And then he joked: "We even take our memory-inducing photos seriously." (Reza in a phone call with Maliheh, March 2021)

Humans transform their smartphones into a device to match their personality and the understanding they have of themselves. This individual work of processing, including, for example, the saving of data, letters, emails, messages, photos and videos, and the downloading of apps, music, etc., makes the smartphone capable of mediating an astonishing depth in terms of intimacy and act as an aid to memory.

A phone's display enables a first look into the personal composition of an indi-

Abb. 11: »Dies ist meine Tochter. Sie ist noch in Syrien, aber wir sprechen jeden Morgen, jeden Abend und jede Nacht«, 2015. ©Grey Hutton

Image 11: "This is my daughter. She's still in Syria, but we speak every morning, evening and night," 2015. ©Grey Hutton

Tiefe an Intimität und Erinnerungshilfe zu vermitteln.

Das Display des Smartphones gewährt einen ersten Blick in die persönliche Gestaltung z.B. in Form des Hintergrundbilds. In einer Fotoreihe für das Online-Magazin *Vice*[10] wählte Grey Hutton die Hintergrundbilder auf den Handys Geflüchteter als Zugang zu deren Geschichten aus (ein Beispiel in Abbildung 11). Während bei manchen Geflüchteten der Nutzen des Smartphones im Vordergrund stand, wiesen viele von ihnen auf die Erinnerungen hin, die sie durch das Handy mit sich tragen konnten – Erinnerungen an Freunde und Familie, die zurückgelassen wurden, an lebende und verstorbene Personen oder auch an Popstars aus der Heimat, deren Musik zu diesem Zeitpunkt voller Bedeutung für sie war.

vidual's phone, in the form of a background photo, for example. In a series of photos taken for the online magazine *Vice*[9], Grey Hutton selected the background photos on refugees' mobile phones as a way of accessing their stories – Image 11 shows one such example. While the functionality of the smartphone was its most important aspect for some of the refugees, many of the refugees also mention the memories that they carry with them through the smartphone – memories of friends and family who were left behind, of other people, both living and dead, and of popstars in their home countries, whose music meant a lot to the phones' owners at this point in their lives.

Abb. 12: Syrische Geflüchtete machen ein Selfie, nachdem sie – nach der Überquerung eines Teils der Ägäischen See von der Türkei aus in einem Schlauchboot – am Strand der griechischen Insel Lesbos angekommen sind, 18. September 2015. ©Yannis Behrakis/Reuters

Image 12: Syrian refugees taking a selfie after arriving on the Greek island of Lesbos, following their crossing of the Aegean Sea from Turkey in a rubber dinghy. 18 September 2015. ©Yannis Behrakis/Reuters

»Selfie« als repräsentatives Tagebuch – und Topos

The Selfie as a Representative Diary – and as Topos

Das Selfie – die digitale Selbstporträtfotografie – ist untrennbar mit dem Smartphone verbunden und ebenso global (vgl. auch Hägele und Schühle 2021). Täglich werden weltweit Millionen von Selfies gemacht, verschickt und über Soziale Medien geteilt. So stellte Google für das Jahr 2014 fest, dass allein Android-Nutzer:innen 93 Millionen Selfies pro Tag machten.[11] 2013 wurde »Selfie« als Wort des Jahres in das Oxford English Dictionary aufgenommen.

Angekommen (Abbildung 12): Nach der erfolgreichen Überquerung des Mittelmeers in einem Schlauchboot von der Türkei aus sind diese jungen Leute sicher am Strand von

The selfie – digital self-portrait photography – cannot be separated from the smartphone as a whole, and is every bit as global as the device is (cf. also Hägele and Schühle 2021). Millions of selfies are taken and sent and shared – by social media every day. Monitoring digital behaviour in 2014, Google concluded that Android users alone take 93 million selfies per day.[10] And in 2013, the Oxford English Dictionary crowned 'selfie' as 'Word of the Year'.

Arrival (Image 12): After successfully crossing the Mediterranean Sea in a rubber dinghy from Turkey, these young adults have safely arrived on shore on Lesbos –

Lesbos angekommen – Grund genug, diesen Moment der Erleichterung und Freude in einem Selfie festzuhalten. Dabei kann ein Selfie-Stick als wichtige Ergänzung zum Smartphone für einen vorteilhaften Winkel sorgen und das Selfie eine Bildsprache aufgreifen, die sich in den Sozialen Netzwerken etabliert und global verbreitet hat. Die selbstbewusste Verwendung dieser Bildsprache, selbst in so existenziellen Situationen wie denen von Flucht und Migration, kann so gelesen werden, dass sich die Menschen mit dieser Art der Dokumentation – trotz aller existenzieller Widrigkeiten und Marginalisierungserfahrungen – der Zugehörigkeit zu den globalen Social-Media-Welten versichern möchten.

reason enough to record this moment of relief and joy in a selfie. In such situations, a selfie-stick can act as an important accessory to the smartphone, providing an advantageous angle and allowing the selfie to address a form of pictorial language that has established itself in a global fashion. The self-confident utilization of this pictorial language, even in a situation as existential as that of flight and migration, can be interpreted as meaning that people want to use this type of visual record to ensure a sense of belonging to worldwide social media worlds, despite all existential adversities and experiences of marginalisation that they have to continually face.

Teilnehmende Smartphones – Autodokumentarfilm

Participatory Smartphones – 'Auto-Documentaries'

Die aufrüttelnde Unmittelbarkeit des 2019 erschienenen Dokumentarfilms *Midnight Traveler*[12] von Hassan Fazili wurde in Filmkritiken hervorgehoben.[13] Der Film wurde ausschließlich mit den drei Smartphones der Familie Fazili gefilmt und zeigt die gefährliche und zermürbende Odyssee von Afghanistan durch Tadschikistan, Iran, Türkei, Bulgarien, Serbien und Ungarn auf der Suche nach Zuflucht in der Europäischen Union. Hassan Fazili und seine Frau Fatima Hossaini, beide bereits Filmemacher:innen in Afghanistan, brachen 2015 mit ihren Töchtern Nargis (damals 11 Jahre alt) und Zahra (6 Jahre alt) auf. Grund ihrer Flucht war Hassan Fazilis TV-Dokumentation über einen abtrünnigen Kommandanten der Taliban. Nachdem diese ausgestrahlt wurde, erhielt er Todesdrohun-

Film critics[11] drew attention to the evocative immediacy of the documentary *Midnight Traveler*, directed by Hassan Fazili, which first screened in 2019.[12] The film was shot exclusively using three smartphones belonging to the Fazili family and shows the dangerous and onerous odyssey from Afghanistan, through Tajikistan, Iran, Turkey, Bulgaria, Serbia and Hungary, in search of refuge in the European Union. Hassan Fazili and his wife Fatima Hossaini, both filmmakers before their journey of flight, left Afghanistan in 2015 with their daughters Nargis and Zahra, who were 11 and 6 years old at the time. It was Hassan Fazili's TV documentary about a renegade Taliban commander that caused them to flee in the first place: Hassan had received death

gen. Unterwegs drohten ihnen Gewalt und Vergewaltigung, Diebstahl und Aussetzung durch ihre skrupellosen Schlepper sowie rassistische Übergriffe in Bulgarien. Nach drei Jahren endete ihre Reise und der Film vorläufig in Ungarn.

Fazili schickte der in den USA lebenden und Persisch sprechenden Filmemacherin und Produzentin Emelie Mahdavian von unterwegs immer wieder aktuelle Handyaufnahmen. Während Fazili noch floh, bemühte sich Mahdavian bereits um die Filmförderung. Das Filmen scheint wohl auch ein vitalisierendes Projekt gewesen zu sein, um die Strapazen bewältigen und bei klarem Verstand bleiben zu können, vor allem während der langen Zeit der Untätigkeit im Flüchtlingslager (in einem serbischen Lager waren sie 475 Tage lang festgesetzt). Der Film vermittelt schlaglichtartige

threats after it was broadcast. Once out of the country and en route to the EU, they were threatened with rape and other forms of violence, had to deal with being robbed and with being abandoned by their unscrupulous traffickers, and were exposed to racist attacks in Bulgaria. After three years on the road, both their journey and the film concluded in Hungary – at least for the time being.

While fleeing, Fazili regularly sent the US-based and Persian-speaking filmmaker and producer Emelie Mahdavian updated mobile phone footage. While Fazili was still in transit, Mahdavian was already working on ways to finance the film. Filming also seems to have been a revitalising project, a way to cope with the hardships and stay sane, especially during the long period of

Abb. 13: Szenenbild aus dem Film *Midnight Traveler*, 2019.
©Oscilloscope Laboratories

Image 13: Scene from the Film *Midnight Traveler*, 2019.
©Oscilloscope Laboratories

Abb. 14: Szenenbild aus dem Film *Midnight Traveler*, 2019.
©Oscilloscope Laboratories

Image 14: Scene from the Film *Midnight Traveler*, 2019.
©Oscilloscope Laboratories

Impressionen einer erzwungenen Migration. Bedrückende, düstere Momente stehen neben manchmal auch heiteren Szenen aus dem Alltag: das atemlose Hasten über eine grüne Grenze, frostige Nächte im Wald und in einer Bauruine, endloses Warten in einer Lagerbaracke, Fatimas Versuche, Fahrradfahren zu lernen, Nargis' coole Tanzbewegungen zur Musik aus dem Smartphone.

inactivity in the refugee camp – they were detained in a Serbian camp for 475 days. The film spotlights particular impressions during a forced migration. Oppressive and bleak moments are juxtaposed with sometimes cheerful scenes from everyday life: the breathless rush across a green border, frosty nights in the forest and in a ruined building, waiting endlessly in a camp barracks, Fatima trying to learn how to ride a bike and Nargis' cool dance moves to music from her smartphone.

Überlebenswerkzeug und mobile Heimat

Das Smartphone ist vieles, wie die Bilder und Geschichten gezeigt haben, insbesondere im Kontext von Flucht und Migration. Es ermöglicht das Finden von Fluchtwegen, die Navigation auf Fluchtrouten, das In-Kontakt-Bleiben mit den Nächsten, die Dokumentation und Kommunikation der eigenen Erlebnisse. Allerdings erzählen die Bilder und Geschichten mehr als von den Funktionen eines technischen Gerätes. Sie zeigen uns Einblicke in die Vergangenheit der Nutzer:innen, in ihr Selbstverständnis, in ihre Fähigkeiten als Smartphone-Reparateur oder Dokumentarfilmer:innen.

Das Smartphone ist mittlerweile für viele Menschen eine Art Zuhause geworden, besonders dann, wenn es kein anderes stabiles Zuhause gibt. Es vereint in sich vieles, was auch das ›klassische‹ Zuhause ausmacht: Es speichert Erinnerungen, gibt Raum für die Selbstentfaltung, für Gespräche mit Familie und Freund:innen. Das Smartphone kann somit als ein ›Ort‹ verstanden werden, in dem wir leben und den wir, ähnlich wie eine Schnecke ihr Haus, mit uns tragen (Miller et al. 2021: 219).

Survival Tool and a Mobile Sense of Home

As the images and stories have illustrated, the smartphone is many things at once, especially in the context of flight and migration. It makes it possible to find and navigate routes of flight, to stay in touch with those you are close to, and to document and communicate individual experiences. That said, the pictures and stories that come to us from these machines narrate more than merely the functions of a technical device. They give us insights into their users' pasts, into their self-images and into their capabilities as smartphone repairers or as documentary filmmakers.

The smartphone has become a kind of home for millions, or even billions, of people, and this is even more the case for those with no other stable home. It brings together many things that constitute what people think of as a 'classic' home: It stores memories and provides space for self-development and for conversations with family and friends. Thus, the smartphone can be understood as a 'place' in which we live and which we carry with us, much like a snail carries its house upon its back (Miller et al. 2021: 219).

Literatur | References

Awad, Isabel; Jonathan Tossell. 2021. Is the Smartphone Always a Smart Choice? Against the Utilitarian View of the ›Connected Migrant‹. *Information, Communication & Society* 24(4): 611–626.

BBC News Africa. 2014. Photo from Djibouti wins World Press Photo Award. BBC News, 14.2.2014, http://www.bbc.co.uk/news/world-africa-26195435

Eriksen, Thomas Hylland. 2020. Filling the Apps. The Smartphone, Time and the Refugee. In: Christine M. Jacobsen, Marry-Anne Karlsen, Shahram Khosravi: *Waiting and the Temporalities of Irregular Migration*, 57-72. London.

Gillespie, Marie. 2016a. ›Smart Migration?‹ The Power and Potential of Mobile Phones and Social Media to Transform Refugee Experiences. *Humanitarian Aid on the Move* 18, 14-19.

Gillespie, Marie. 2016b. »Phones – Crucial to Survival for Refugees on the Perilous Route to Europe«. *The Conversation.com*, 16 May 2016. https://theconversation.com/phones-crucial-to-survival-for-refugees-on-the-perilous-route-to-europe-59428

Gillespie, Marie; Lawrence Ampofo; Margaret Cheesman; Becky Faith; Evgenia Iliadou; Ali Issa; Souad Osseiran; Dimitris Skleparis. 2016. Mapping Refugee Media Journeys. Smartphones and Social Media Networks. https://www.open.ac.uk/ccig/sites/www.open.ac.uk.ccig/files/Mapping%20Refugee%20Media%20Journeys%2016%20May%20FIN%20MG.pdf

Gillespie, Marie; Souad Osseiran, Margie Cheesman. 2018. Syrian Refugees and the Digital Passage to Europe: Smartphone Infrastructures and Affordances. *Social Media + Society* 4(1): 1–12.

Hage, Ghassan. 2009. *Waiting*. Melbourne.

Hägele, Ulrich; Judith Schühle. 2021. *SnAppShots: Smartphones als Kamera*. Münster.

Lauser, Andrea; Antonie Fuhse; Peter J. Bräunlein; Friedemann Yi-Neumann. 2022. Introduction: From ›Bare Life‹ to ›Moving Things‹: On the Materiality of (Forced) Migration. In: Friedemann Yi-Neumann, Andrea Lauser, Antonie Fuhse, Peter. J. Bräunlein: *Material Culture and (Forced) Migration. Materializing the Transient*. London.

Miller, Daniel; Laila Abed Rabho; Patrick Awondo; Maya de Vries; Marília Duque; Pauline Garvey; Laura Haapio-Kirk; Charlotte Hawkins; Alfonso Otaegui; Shireen Walton; Xinyuan Wang. 2021. *The Global Smartphone. Beyond a Youth Technology*. London.

Miller, Daniel. 2021. A Theory of a Theory of the Smartphone. *International Journal of Cultural Studies* 24(5): 860–876.

Ponzanesi, Sandra; Koen Leurs. 2014. On Digital Crossings in Europe. *Crossings: Journal of Migration and Culture* 5 (1): 3–22.

Turner, Simon. 2015. ›We Wait for Miracles‹ – Ideas of Hope and Future Among Clandestine Burundian Refugees in Nairobi. In: Elizabeth Cooper, David Pratten: *Ethnographies of Uncertainty in Africa*, 173–192. London.

Anmerkungen

1 Die Geschichten sind inspiriert von und basieren auf eigenen Feldforschungsbegegnungen und -daten, besonders von Maliheh Bayat Tork, aber auch auf Recherchen in weltweit verfügbaren Medien. Die existenzielle Bedeutung von Smartphones in der Gestaltung forcierter Migration ist in vielfältigen Medienformaten dokumentiert und illustriert worden, wie zum Beispiel in Online-Ausstellungen wie *Connected Refugees* des Künstlers Ai Weiwei (http://artisticlab.forumviesmobiles.org/en/the-refugee-project) oder in Dokumentarfilmen wie *#myescape*, in dem viele Selbstdokumentationen zu einer Collage zusammengeschnitten wurden (zum Film siehe auch https://www.dw.com/de/tv/myescape/s-32603), und *District Zero* (https://www.districtzero.org/; mehr zu diesem Film im nachfolgenden Abschnitt.

2 Für den Trailer zum Film *Disctrict Zero* siehe die offizielle Website siehe https://www.districtzero.org/

3 Jean Dykstra: »John Stanmeyer wins World Press Photo Award«, *photograph*, 15 Februar 2014, https://photographmag.com/articles/john-stanmeyer-wins-world-press-photo-award/. Übersetzung: Antonie Fuhse.

4 Übersetzung: Antonie Fuhse.

5 Laut Marie Gillespie (2016) haben selbst die ärmsten Menschen auf der Flucht in der Regel Zugang zu einem Mobiltelefon, wenn auch nicht zu einem hochmodernen Modell.

6 Marion MacGregor: »Smartphone Apps Helping Migrants to Find Local Services«, *InfoMigrants*, 29. Juni 2018, https://www.infomigrants.net/en/post/10096/smartphone-apps-helping-migrants-to-find-local-services

7 Siehe auch den Beitrag von Romm Lewkowicz in diesem Band.

8 Siehe: BBC Media Action »Your Phone Is Now a Refugee's Phone [watch on a mobile]«, *BBC*, 27. September 2018, https://www.youtube.com/watch?v=m1BLsySgsHM

9 Entstanden ist dieser Kurzfilm in Zusammenarbeit von BBC Media Action und der humanitären

Notes

1 The stories are inspired by, and based on, encounters documented and data gathered during the authors' own field research, particularly that of Maliheh Bayat Tork, but also on research conducted through globally available forms of media. The existential significance of smartphones in the formation of forced migration has been documented and illustrated in heterogeneous media formats, including, for example, online exhibitions like *Connected Refugees* curated by the artist Ai Weiwei (last accessed 22 November 2021: http://artisticlab.forumviesmobiles.org/en/the-refugee-project), or in documentary films like *#myescape*, in which films in which many individuals had documented themselves were compiled to form a visual collage. On the film see also, last accessed 22 November, 2021: https://www.dw.com/de/tv/myescape/s-32603), and *District Zero*, last accessed 22 November 2021: https://www.districtzero.org/; for more on the latter film, see the following section.

2 To see the trailer for the film *District Zero*, see the film's official website. Last accessed 23 November, 2021: https://www.districtzero.org/

3 Jean Dykstra: "John Stanmeyer wins World Press Photo Award," *photograph*, 15 February, 2014, https://photographmag.com/articles/john-stanmeyer-wins-world-press-photo-award/

4 According to Marie Gillespie (2016), even the poorest people who flee normally have access to some kind of mobile phone, even if they don't have access to a highly-modern model.

5 Marion MacGregor, "Smartphone Apps Helping Migrants to Find Local Services", *InfoMigrants*, 26 June 2018.

6 See also the article by Romm Lewkowicz in this volume.

7 See the BBC media action: "Your Phone Is Now a Refugee's Phone [watch on a mobile]", published on 18 July 2016 at https://www.youtube.com/watch?v=m1BLsySgsHM

8 This short film took shape as a result of a cooperation between BBC Media Action and the humanitarian organisation DAHILA in the context of a study aimed at analysing communication behaviour and information requirements pertaining to refugees during their journeys of flight. The analysis was carried out in transit camps in Greece and in Germany. The film in intended to demonstrate the importance of

Organisation DAHILA im Rahmen einer Studie mit dem Ziel, Kommunikationsverhalten und Informationsbedarf von Flüchtenden auf ihrer Reise, in Transit-Lagern in Griechenland sowie in Deutschland zu analysieren. Der Film soll die Wichtigkeit von Kommunikation und Information in humanitären Krisen zeigen, und ist inspiriert von Geschichten aus der Studie. Für weitere Informationen siehe BBC Media Action: »Voices of Refugees«, http://bbc.in/2amioOP (Zugriff am 8.7.2021).

10 Siehe Grey Hutton, Barbara Dabrowska und Yahya Al: »We Asked Some Refugees for the Stories Behind Their Smartphone Backgrounds«, *Vice*, 17. September 2015 (https://www.vice.com/en/article/avynke/the-smartphones-of-refugees-876).

11 Siehe Richardt Brandt: »Google Divulges Numbers At I/O: 20 Billion Texts, 93 Million Selfies and More«, *Silicon Valley Business Journal*, 25. Juni 2014 (https://www.bizjournals.com/sanjose/news/2014/06/25/google-divulges-numbers-at-i-o-20-billion-texts-93.html).

12 Für die Website und den Trailer von *Midnight Traveler* siehe https://midnighttraveler.oscilloscope.net/

13 Hier sei u.a. auf die Filmkritik von Stefan Volk im *filmbulletin* vom 9.12.2019 (https://www.filmbulletin.ch/full/filmkritik/2019-12-9_midnight-traveler/) und die Kritik von Kenneth Turan in der *Los Angeles Times* vom 3.10.2019 (https://www.latimes.com/entertainment-arts/movies/story/2019-10-03/midnight-traveler) verwiesen.

communication and information in humanitarian crises, and is inspired by stories from the research project. For further information, see BBC Media Action, "Voices of Refugees", last accessed on 8 July 2021: http://bbc.in/2amioOP

9 See: Grey Hutton, Barbara Dabrowska and Yahya Al: "We Asked Some Refugees for the Stories Behind Their Smartphone Backgrounds", *Vice*, 17 September 2015 (https://www.vice.com/en/article/avynke/the-smartphones-of-refugees-876).

10 See Richardt Brandt: "Google Divulges Numbers at I/O: 20 Billion Texts, 93 Million Selfies and More", *Silicon Valley Business Journal*, 25 June 2014, https://www.bizjournals.com/sanjose/news/2014/06/25/google-divulges-numbers-at-i-o-20-billion-texts-93.html

11 See, among others, the review by Stefan Volk in *filmbulletin* on 9 December 2019, https://www.filmbulletin.ch/full/filmkritik/2019-12-9_midnight-traveler/) and the review by Kenneth Turan in der *Los Angeles Times* on 3 October 2019, https://www.latimes.com/entertainment-arts/movies/story/2019-10-03/midnight-traveler

12 For the website and trailer for *Midnight Traveler*, go to https://midnighttraveler.oscilloscope.net/

Die Objektivierung von Migration in den Darstellenden Künsten

Eine Annäherung des freien Theaters »boat people projekt« aus Göttingen

The Objectivisation of Migration in the Performing Arts

A Joint Publication from the Independent Theatre Company "boat people projekt" from Göttingen, Germany

Abb. 1: Foto von der Preisverleihung Theaterpreis des Bundes, 2019. ©Eva Radünzel

Image 1: Photo from the award ceremony of the Federal Theatre Prize, 2019. ©Eva Radünzel

Gold –
Die Künstlerin ist abwesend
von Luise Rist

> Wo ist mein Rock aus allen Blumen der Welt?
> Vor langer Zeit
> Nähte ich ihn mir.
>
> *Papusza*

Gewidmet der polnischen Dichterin Bronisława Papusza Wajs, die von 1910 bis 1987 lebte.

Rollen:
Irina Baryalei: Bildende Künstlerin, Ende 30 oder älter, bevorzugt zu besetzen mit einer Person of Colour (POC).
Papusza: Dichterin, jedes Alter möglich, bevorzugt zu besetzen mit einer Romnija, in jedem Fall mit einer POC.
Danyar: Taxifahrer und Chemiker, jedes Alter möglich, Schwarz.
Stimme: Teil der fiktiven Vernissage ist eine weibliche Stimme auf Romanès. Diese Texte sollten unbedingt von einer Romanès-Muttersprachlerin eingelesen werden. Das hier verwendete, in Polen gesprochene Romanès kann von der Muttersprachlerin in das von ihr gesprochene Romanès übertragen werden.

Eine Kunstausstellung. Die Bühne ist der Ort einer fiktiven Vernissage. Der Bühnenraum ist ein Ausstellungsraum. Die ausgestellten Objekte sind inspiriert von der hinduistischen Göttin Kali sowie der christlichen Schwarzen Madonna Sarah La Kali.

Stimme im Hintergrund, wiederkehrend.

STIMME: *Kaj san?*

Gold –
The artist is absent
by Luise Rist

> Where is my dress, sewn from all the world's flowers?
> I sewed it
> A long time ago.
>
> *Papusza*

Dedicated to the Polish poet Bronisława Papusza Wajs, who lived from 1910 to 1987.

Roles:
Irina Baryalei: A visual artist, in her late 30s or older, to be cast, preferably, with a Person of Colour.
Papusza: Poet, can be of any age, preferably to be cast with a Roma woman, in any case with a PoC.
Danyar: Male taxi driver and chemist, can be of any age, black.
Voice: One part of the vernissage is a female voice speaking Romani. These texts should definitely be recorded by a female native speaker of Romani. The form of Romani used here, the form spoken in Poland, can be transposed by the female native speaker into the type of Romani that she speaks.

An art exhibition. The stage is the location for a fictional vernissage. The stage space is an exhibition space. The objects exhibited are inspired by the Hindu goddess Kali and the Christian Black Madonna, Sarah la Kali.

A voice that repeats in the background.

VOICE: *Kaj san?*

IRINA: Guten Abend. *Kaj san*, hören Sie jemanden fragen. Das bedeutet: Wo bist du?

Wo bist du ...

Im Katalog dieser Ausstellungsreihe mit dem Titel GOLD, die mehrere neue – ... Kunst von Frauen präsentiert, werde ich als »Performerin und Bildende Künstlerin Irina Baryalei« angekündigt. Zur heutigen Eröffnung wurde ich gebeten, Ihnen einen »tiefen Einblick« in meinen kreativen Prozess zu geben, Sie teilhaben zu lassen an meiner Auseinandersetzung mit dem Thema. Was ist das Thema? Die Ausstellung ist einer Dichterin der Roma gewidmet, deren Name Ihnen voraussichtlich kein Begriff ist. Beschäftigt habe ich mich mit dieser Dichterin, ja, und mit Erinnerung – mit kollektiver Erinnerung. Oder besser: kollektivem Ignorieren ... Thema dieser Kunstobjekte ist aber nicht Flucht und Vertreibung. Das Thema ist die Ausstellung selbst. Ausstellen und Ausgestelltwerden. Wer zeigt wem etwas von sich selbst, wo beginnt meine Grenze, wo löst sie sich auf? Um Ihnen dies spürbarer zu machen, habe ich etwas vorbereitet.

Sobald bei Ihnen am Platz ein Lämpchen aufleuchtet, möchte ich Sie bitten, den Satz, der dort steht, vorzulesen. Satz 1. Wenn das Lämpchen später wieder aufleuchtet, lesen Sie bitte Satz 2. Die Sätze habe ich Ihnen in den Mund gelegt. Bitte verzeihen Sie, wenn manche Sätze nicht angenehm auszusprechen sind. Es sind nicht Ihre eigenen Sätze. Jeder Satz spielt eine Rolle. Es ist NICHT Ihre Rolle.

Können wir starten? Technik? Lampe 1 bitte.

So beginnt das Theaterstück Gold, das von Luise Rist, einer der Mitgründerinnen des freien

IRINA: Good evening. The voice you're hearing is asking: *kay san*, meaning: Where are you?

Where are you ...

In the catalogue of the exhibition series, titled GOLD, which presents a number of new ... pieces of art by women, I was billed as "Female Performer and Visual Artist, Irina Baryalei". I was also asked to give the public a "deeper insight" into my creative process for the opening of the exhibition today, to enable them to participate in my engagement with the theme of the exhibition. But what is the theme? This exhibition is dedicated to a female Roma poet whose name presumably means nothing to most of you. Yes, I have engaged with this poet and with memory – with collective memory. Or to express that slightly better: with collective and intentional ignoring ... the theme of these art objects is not flight or expulsion. Instead, the theme is the exhibition, and exhibiting itself. Exhibit – and be exhibited! Who shows whom something of themselves, where does my boundary begin, and where does it dissolve? I have prepared something to make these questions tangible.

As soon as a little lamp lights up beside your seat, I would like to ask you to read the sentence that is shown on that lit-up screen. Sentence 1. If the lamp lights up again, later, please read sentence 2. These are words and sentences that I have put into your mouth. Please forgive me if some of the sentences are not pleasant to pronounce. They are not your own sentences. Each sentence has a part to play. This is NOT your part.

Theaters »boat people projekt« in Göttingen, geschrieben wurde. Im vorliegenden Text wechseln sich Auszüge aus dem Theaterstück mit Aussagen von Künstler:innen aus dem Team des »boat people projekts« ab. Die Künstler:innen wurden von Anoush Masoudi, Mitarbeiter im Projekt »Zur Materialität von Flucht und Migration«, zum Themenkomplex Theater und Objektivierung von Migration befragt. Er fragte danach, inwieweit Erinnerungen Objekte für die künstlerische Arbeit sein können und welche Rolle Einwanderungsgeschichte für die Theaterarbeit spielt. Die Fragen führten zu einem fruchtbaren Austausch über Kategorisierungen und Zuschreibungen. Bevor wir zu den Antworten der Künstler:innen kommen, führt uns der Text des Theaterstücks an zentrale Debatten zum Thema Ausstellen heran.

1: Können Sie uns etwas über diese Dichterin erzählen, die Sie erwähnt haben, und über diese, ich glaube, indische Figur Kali?
IRINA: Papusza, auf Polnisch Bronisława Wajs genannt, war eine Dichterin, obwohl sie ohne direkten Zugang zu Schriftsprache aufwuchs. Ihre große Familie bestand aus Musikern, aus fahrendem Volk. Niemand von ihnen war je zur Schule gegangen. Ihre Musik wurde von Generation zu Generation weitergegeben, und blieb immer bei ihnen. Die Roma und Sinti wurden im Holocaust, der bei ihnen Porajmos heißt – das große Verschlingen – zu Hunderttausenden ermordet. Papusza war eine Überlebende. Über ihre Kindheit in den Wäldern Polens, über die Vertreibung durch die Nazis, über den Mord an 500.000 Roma und Sinti hat sie geschrieben. Aber ihre Leute wollten nicht beschrieben werden. Papusza hat das Schreiben tief bereut. Die Spur der Roma führte mich nach Indien.

Can we start? Stage technicians ready? Light One, please.

This is how the play Gold begins, written by Luise Rist, one of the women who co-founded the independent theatre company "boat people projekt" in Göttingen. In the following, excerpts from Rist's play alternate with statements made by artists from the "boat people projekt" team. The artists were interviewed by Anoush Masoudi, a staff member for the research project "On the Materiality of (Forced) Migration" ["Zur Materialität von Flucht und Migration"], about the complex and intersecting themes of theatre and the objectivisation of migration. Masoudi's questions explore the extent to which memories can also be objects to be used in artistic labour, and investigate which roles the history of immigration plays in theatre work. The questions led to a fruitful exchange regarding categorisation and ascription. Before we listen to the artists' answers to these questions, the text of the theatre play brings us nearer to crucial debates on the subject of exhibiting.

1: Can you tell us something about this female poet that you mentioned, and something about this figure Kali who I believe, is Indian?
IRINA: Papusza – called Bronisława Wajs in Polish – became a poet, despite having grown up without much access to written language. Her large family was made up of musicians, of travelling folk. None of them had ever been to school. Their mu-

Auf meinen Reisen begegnete mir die hinduistische Göttin Kali, die Schwarze Kali. Ich habe angefangen, Kali-Tempel zu besuchen, ich habe Kali immer wieder gemalt, ich war fasziniert von dieser weiblichen schwarzen Gottheit, die als Göttin der Roma gilt.
STIMME: *Kaj san?*
　Ja, bitte?
2: Jede Ausstellung stellt Kunst aus. Das liegt in der Natur der Sache. Wieso ist das problematisch? Sie wollen doch etwas zeigen. Sonst würden Sie keine Kunst machen. Mir kommt das etwas esoterisch vor. Auch, dass ich hier gerade etwas ablese.
3: Lassen Sie sich doch darauf ein.
2: Ist das eine Psychotherapie?
IRINA: Solange die Therapeutin die Göttin Kali ist … bin ich einverstanden …
　Was ist problematisch im Umgang mit, sagen wir, einer anderen Kultur. Sehen Sie die Schwarze Madonna. Sara La Kali. Das Objekt, das Sie hier sehen, ist einer Statue nachempfunden, die ich in Südfrankreich sah. In Saintes Marie de la Mer wird seit mehr als 2.000 Jahren eine besondere Statue angebetet; sie trug im Lauf der Zeiten wohl immer andere Namen. Die Figur war immer weiblich und hatte eine schwarze Hautfarbe. Es ist eine Kali-Figur. Sara la Kali entspringt ganz genau der gleichen Wurzel wie die weibliche, schwarze Gottheit Kali in Indien. Die nur viel extremer aussieht mit ihren vielen Armen und der ausgestreckten Zunge. Es gibt eine uralte Geschichte der Anbetung einer weiblichen und eben auch schwarzen Gottheit, die weltweit fast identisch existiert, aber verschiedene Namen trägt, und – das ist entscheidend: die von den Gesellschaften marginalisiert wurde. Interessanterweise wird Kali sowohl in Indien

sic was handed down from generation to generation and always remained in their possession. Then, hundreds of thousands of Roma and Sinti were murdered in the Holocaust, which is called the *Porajmos* [the Devouring]. Papusza was a survivor. She wrote about her childhood in Poland's forests, about the expulsion by the Nazis and about the murder of 500,000 Roma and Sinti. But her people did not want to be described. Papusza deeply regretted writing. Following in the Roma's footsteps also led me to India.
　On my travels, I encountered the Hindu goddess Kali, Black Kali. I started to visit temples to Kali, I painted Kali repeatedly and I became fascinated by this black, female goddess, who is also seen as the Roma's goddess.
VOICE: *Kaj san?*
　Yes, how can I help?
2: Every exhibition means exhibiting art. That's the nature of the beast. What's problematic about that? You want to show something, don't you, otherwise you wouldn't make art? That seems rather esoteric to me. As does the fact that, right now, I'm reading aloud a prepared text.
3: Try and be open for it.
2: Is this a kind of psychotherapy?
IRINA: As long as the therapist is the goddess Kali then … that's OK with me …
　What's problematic about, say, engaging with a different culture. Look at the Black Madonna, Sara la Kali. The object that you see is based on a statue that I saw in the south of France. This unusual statue in Saintes Marie de la Mer has been prayed to for over 2,000 years: Her name

als auch in Kanada, oder Mexiko, oder eben in Saintes Marie de la Mer insbesondere von Roma und Sinti verehrt. --

Wenn ich eine Kali-Figur reproduziere, riskiere ich nicht nur, dass sie ohne den Kontext bedeutungslos wird – denken Sie an tausendfach reproduzierte Buddha-Statuen, die in deutschen Wohnzimmern sitzen –, sondern ich riskiere auch, dass ich mich einer Kultur bediene, dass ich kaufe, ohne zu bezahlen. Verstehen Sie, was ich meine? – Das ist ein schmaler Grat. Ich möchte eine Figur natürlich nicht kopieren, sondern in einem anderen Kontext mit einer neuen Bedeutung aufladen. Das ist übrigens nicht esoterisch. Sondern sehr politisch.

Es kann passieren, dass ich etwas von einer anderen Kultur ausstelle und dadurch der Kultur nicht etwas hinzufüge, sondern ihr etwas zufüge. Eine Silbe nur ist anders, aber …
3: Sie sprechen von kultureller Aneignung.
IRINA: Ja. Das Ausgestellte kann aber auch Teil meiner eigenen Kultur sein. Auch dann gibt es den Moment der Gefährdung – was zeige ich, was stelle ich aus, ohne jemanden persönlich auszustellen, ohne der Community etwas zu nehmen, was sie vielleicht nur für sich bewahren möchte, bewahren muss? Die Dichterin Papusza, deren Stimme Sie in meiner Installation hören können, ist so ein Beispiel. Ein polnischer Intellektueller brachte sie dazu – auch seine Beweggründe wären zu beleuchten! –, ihre Gedichte und Lieder zu veröffentlichen. Durch die Veröffentlichungen wurde ein Teil der Kultur der polnischen Roma für alle zugänglich. Für die Gadjé, so heißen die »Nicht-Roma«. Für die Roma war es nicht zu ertragen, dass etwas von ihnen, aus ihrer Kultur, zu den Gadjé gekommen war. Die ihnen seit jeher nur schaden wollten. Es war

has changed on innumerable occasions over the centuries. But the figure was always female and always had black skin. It is a Kali figure. Sara la Kali has exactly the same roots as the black, female deity Kali in India. The latter manifestation of the deity simply looks more extreme, with her many arms, and her tongue stuck out. There is an ancient story of praying to a goddess – who's also black – which exists in an almost identical form worldwide, but who bears different names and who – this is decisive – is marginalised in the respective societies where it manifests itself. It is interesting that Kali is worshipped in India as in Canada or in Mexico or in Saintes Marie de la Mer, where primarily Roma and Sinti revere her.

So, when I reproduce a Kali figure, I risk not only that it will become meaningless when deprived of the context these figures normally reside in – think just of the thousands of Buddha reproductions in German sitting-rooms – but also that I help myself to a culture, bits of which I buy up, without actually paying for it. Can you follow what I mean? It's a fine line. I obviously don't want to simply copy a figure but rather to charge it with new significance, in a different context. As it happens, that is not esoteric. But rather very political.

It's possible that, in choosing to exhibit something from a different culture, I don't add anything enriching to that culture, but simply give it another blow to the head.
3: You're talking about cultural appropriation?
IRINA: Yes. But what is exhibited can also be a part of my own culture. There is a

Die Objektivierung von Migration

Abb. 2: Javeh Asefdjah, Sandrinne Ugitoh Essem, Mark Kutah in der Produktion GOLD, 2020. @Luise Rist

Image 2: Javeh Asefdjah, Sandrinne Ugitoh Essem, Mark Kutah in the production of GOLD, 2020. @Luise Rist

Verrat in ihren Augen. Im Angesicht des Holocaust nur zu verständlich. Und dennoch ...
4: Ist das hier Ausgestellte denn nun Teil Ihrer eigenen Kultur?
STIMME: *Jekh chavorri pr-e phike, vaver apr-e vasta. Ada zanäs Rroma.*
IRINA: Gibt es ein technisches Problem? Die Stimme kommt hier eigentlich nicht.
5: Was meinen Sie?
IRINA: Wie, was meinen Sie? Ich? Wo ist Papusza?
TECHNIK: Alles nach Plan, wir können weitermachen.

Irina wirkt abwesend. Danyar versucht, unauffällig Kontakt aufzunehmen.

dangerous moment here, too – what do I show, what do I exhibit, without acting as an exhibitionist regarding other individuals or groups, without taking something away from the community? Something that they might like or might have to keep just for themselves? The poet Papusza, whose voice you can hear in my installation, is an example of this tendency. A Polish intellectual, whose motives ought also to be examined, persuaded her to publish her poems and songs. The published works made a piece of Polish Roma accessible to everyone. Accessible to the *Gadjé*, the Romani word for non-Roma, many found it unbearable that a part of themselves, and a part of their culture, had

The Objectivisation of Migration

DANYAR: Alles in Ordnung?
TECHNIK: Wenn Sie bitte nur sprechen, wenn die Lampe bei Ihnen leuchtet!
DANYAR: Klar. Verzeihung …
6: Sie haben sich sehr mit dem Thema der Minderheit beschäftigt – ich nehme nicht an, dass Sie selbst eine … eine Roma sind, also wie kommen Sie auf das Thema?

Das »boat people projekt«

Nina de la Chevallerie, Theaterregisseurin und Produzentin ist Mitbegründerin des freien Theaters. Sie leitet dazu auch soziokulturelle Gruppen und ist als Netzwerkerin tätig.

Seit 2009 arbeitet das freie Theater »boat people projekt« in Göttingen in verschiedenen Konstellationen schwerpunktmäßig zum Thema Flucht und Migration. Unser Hauptfokus und der inhaltliche Diskurs beschäftigen sich in möglichst radikaler Konsequenz und mit ehrlicher Reflexion mit der Frage, wie wir eine ›vielfaltssensible Öffnung‹ erreichen, was Diversität überhaupt bedeutet, welche Repräsentant:innen bei uns vorhanden sind, wie wir strukturellen Rassismus verhindern und ihm entgegentreten. Wie kann politisches Theater wirksam sein? Welchen Beitrag können wir zur ›Migrationsdebatte‹ leisten?

Die Frage, inwieweit Migration per se ein Objekt oder das Thema für Theaterstücke sein kann und welche Konsequenz dies für die künstlerische Zusammenarbeit hat, ist von Beginn an ein Thema in jedem Vorhaben. Der Impuls, die Arbeit jenseits der gewohnten Stadttheaterarbeit zu beginnen, kam 2009 durch die Schlagzeilen über Lam-

fallen into the hands of the *Gadjé*. The people who had only always wanted to harm them: Polish Roma viewed this as treason. And in the context of the Holocaust, this attitude is only too understandable. Yet still …
4: Is then what is exhibited here simply then a part of your own culture?
VOICE: *Jekh chavorri pr-e phike, vaver apr-e vasta. Ada zanäs Rroma.*
IRINA: Is there some kind of technical problem? The voice doesn't actually come in at this point.
5. What do you mean?
IRINA: What do you mean: "What do you mean?" Me? Where is Papusza?
TECHNICIAN: We're following the plan for everything, we can go on.

Irina seems like she isn't really there. Danyar tries inconspicuously to start up a conversation with her.

DANYAR: Is everything alright?
TECHNICIAN: Could you please only speak when the light beside you switches on!
DANYAR: Of course. Apologies …
6. You have engaged intensively with the theme of minority – I guess that you yourself are not … a Roma, so how did you arrive at this subject?

The "boat people projekt"

Nina de la Chevallerie, theatre director and producer, is one of the co-founders of this independent theatre company, where she also leads sociocultural groups and is involved as a networker.

Die Objektivierung von Migration

pedusa und die vermehrten Schiffsunglücke dort zustande. Luise Rist und ich begannen zu recherchieren, inwieweit auch hier in Göttingen, in Niedersachsen Geflüchtete ›angekommen‹ waren, wie die Lebenssituation der Menschen ist, welche Biographien sich hinter den Schlagzeilen verbergen. Im ersten Stück LAMPEDUSA standen Geflüchtete aus Äthiopien neben Studierenden aus Kamerun und Eingewanderten aus Ghana auf der Bühne – einem Stadtbus, der die mitfahrenden Zuschauenden an vertraut geglaubte Orte in Göttingen brachte.

Sowohl in LAMPEDUSA als auch in den späteren beiden Jahren mit den Stücken MIKILI und KEINSTERNHOTEL (2010 und 2011) haben wir auf Migration und Flucht als Thema fokussiert. In den Schauspielproduktionen bestand das Ensemble jeweils aus professionellen Schauspieler:innen aus Deutschland und der Schweiz sowie geflüchteten, meist jungen Menschen, aus afrikanischen Ländern. Wir bildeten intime, einander zugewandte Gruppen, erfuhren voneinander und übereinander. Wir lernten viel über deutsches Asylrecht und die politische Situation in den Herkunftsländern der neuen Kolleg:innen und machten diese zum Thema der Stücke. Gemeinsam Theater machen über Migration bedeutete damals wie heute auch, dass Grenzen verschwimmen, Professionelles nicht von Privatem zu trennen ist. Der Kontakt mit Ausländerbehörden und Sozialarbeiter:innen blieb nicht aus. Während der Proben wurde in vielen Sprachen diskutiert: Wie können wir komplexe Sachverhalte wie die Gründe für Flucht möglichst ohne Worte ausdrücken, welche Bilder werden von wem auf welche Art gelesen?

Since 2009, the independent theatre "boat people projekt" in Göttingen has been working in various constellations to focus on the topic of flight and migration. Our main focus and the discursive content deal, as radically as possible and using honest reflection, with the questions of how we can arrive at a "diversity-sensitive opening", what diversity actually means, which representatives we have and how we can prevent and counter structural racism. How can political theatre display its agency? Which contributions can we make to the societal conversation on migration?

The question regarding the extent to which migration can, in itself, be an object or theme for theatre plays, and what consequences this has for artistic collaboration, has been an issue in every project from the very beginning. The impulse to start working beyond the reaches of typical, German municipal theatre came in 2009, prompted by the headlines about Lampedusa and the increased number of sea vessels sinking there. Luise Rist and I responded by starting to research the extent to which refugees had also 'arrived' here in Göttingen, in the federal state of Lower Saxony, what the living situations of these people were like, what biographies were hidden behind the headlines. In the first play, LAMPEDUSA, refugees from Ethiopia stood on stage next to Cameroonian students and Ghanaian immigrants. The stage was a city bus, which took the audience to places in Göttingen that they thought they were familiar with.

Both in LAMPEDUSA, and in the two subsequent years in the plays MIKILI and

IRINA: Die Roma und Sinti haben eine reiche Mythologie. Sie sind die ersten Kosmopoliten ...
7: Naja –
IRINA: Wie? Ich bitte Sie, abzulesen, was da steht. Eigene Kommentare ok, aber – das ist eine Performance, die läuft nach bestimmten Regeln, sie können gerne später – ach egal. Sie haben Recht. Es ist natürlich möglich, dass Sie selbst ...
STIMME: *Ada zanäs Rroma.*
IRINA: Ok. Wieso nehmen Sie an, dass ich keine Romnija oder Sintezza bin?
8: Möchten Sie uns denn erzählen, wo Sie selbst herstammen?
9: Ich habe gelesen, dass Sie in Isfahan, im Iran, geboren sind.
IRINA: Roma gibt es auch in Isfahan. Meine Kunstwerke sprechen auch nicht von mir. Sondern von uns. Hoffentlich. Ich bin im Übrigen auch nicht in Isfahan geboren. – Sondern in der Türkei. Mein Vater kam aus – ach.
10: Kamen Ihre Eltern als Gastarbeiter aus der Türkei?
IRINA: Ich bin nur geboren in der Türkei, meine Eltern sind keine Türken. Das ist kompliziert und hat mit dem Thema der Ausstellung auch nichts zu tun.
11: Die Sinti werden auch in der Türkei ausgegrenzt, das habe ich gehört, dass es da große Probleme gibt.
12: Die meisten Roma leben in Rumänien.
IRINA: ... Ich bin in Deutschland aufgewachsen.
13: Stimmt es eigentlich, dass Ihre Familie abgeschoben werden sollte?

Stille

KEINSTERNHOTEL (2010 and 2011, respectively), we concentrated on the subjects of migration and flight. In each of these theatrical productions, the ensemble consisted of professional actors from Germany and Switzerland alongside mostly young refugees from African countries. We formed intimate groups of people that felt reciprocally approachable and learned from and about each other. We gathered much knowledge about German asylum law and the political situation in the new colleagues' countries of origin – and then made these contents the theme of the plays. Making collaborative theatre with migration as a subject meant, then as now, that boundaries become blurred, that the professional cannot be separated from the private. It was impossible to entirely eschew contact with the authorities for non-German nationals[1] and with social workers. Discussions during rehearsals were held in many languages: How can we express complex issues such as the reasons for flight without, as far as possible, using words? Which images are read by whom and in which way?

IRINA: The Roma and Sinti possess rich mythologies. They are the first cosmopolitans ...
7: Not sure about that –
IRINA: What? Would it be OK for you just to read what's in your script? Some of your own comments, all right, but – this is a performance that works according to particular rules, you're welcome later ... – oh, forget it. You're right. Of course, it's possible that you're a ...
VOICE: *Ada zanäs Rroma.*

IRINA: Das steht da nicht.

14: Sie sind Künstlerin geworden. Das ist ermutigend. Die meisten Gastarbeiter, die nach Deutschland kamen, haben ja in Fabriken gearbeitet. Dass deren Kinder dann hier Karriere machen, das ist doch ein Zeichen für gelungene Integration.

15: Gastarbeiter? Sie haben da wohl etwas falsch verstanden. Es ist jedenfalls fantastisch, dass Frau Baryalei jetzt die Gelegenheit hat, hier bei uns auszustellen.

16: Unsere kleine, aber durchaus renommierte Galerie hat einen besonderen Ruf für bildende Künstlerinnen. Wir sind bekannt als Entdecker für neue weibliche Talente. Wir fördern besonders die Kunst von Frauen.

IRINA: Fantastisch.

17: Es ist wirklich schön zu sehen, wie die zweite und die dritte Generation der Einwanderer in Deutschland ankommt. Nicht nur in der Arbeitswelt, sondern auch in der Kunstwelt.

18: Ich möchte das nicht lesen, was hier steht.

1: Wenn man bedenkt, dass es in den Herkunftsländern teilweise kaum Schulbildung gab. Von Kunstunterricht ganz zu schweigen.

Die Künstlerin tritt in Kontakt mit einem ihrer Objekte. Kommentarlos stellt sie eine Kunstperformance vor. Die Performances sind Teil ihrer geplanten Inszenierung.

Migrantische Künstlerin = migrantische Kunst?

Javeh Asefdjah, Schauspielerin. Javeh Asefdjah studierte an der Hochschule für Musik und Theater »Felix Mendelssohn Bartholdy« in Leipzig

IRINA: OK. Then why do you assume that I'm not a Romniya or a *Sintezza* myself?

8: Would you like to then tell us where you come from yourself?

9: I read that you were born in Isfahan in Iran.

IRINA: There are also Romani People in Isfahan. Moreover, my artworks are not primarily about me, but rather about us. That's what I hope for, at least. As it happens, I wasn't born in Isfahan either. I was actually born in Turkey. My father came from … hmm.

10: Did your parents move to Turkey as invited migrant workers?

IRINA: I was just born in Turkey, but my parents aren't Turks. It's complicated – and also has nothing to do with the subject of the exhibition.

11: I heard that the Sinti were also marginalised in Turkey, that there were large problems to deal with there, too.

12: Most Roma live in Romania.

IRINA: … I grew up in Germany.

13: Is it true that your family is to be deported?

Silence

IRINA: That's not in the script.

14: You became a female artist. That's encouraging. Most migrant workers that came to Germany in the 1950s and 1960s worked in factories.

15: Migrant workers? You've definitely got the wrong end of the stick somewhere. Anyway, it's fantastic that Mrs Baryalei now has the opportunity to exhibit here in our space.

16: Our small but unquestionably renowned gallery has a particularly good reputation for its female visual artists. We're known as

Abb. 3: Javeh Asefdjah in der Produktion GOLD, Bühnenbild: Petra Straß, 2020. ©Luise Rist

Image 3: Javeh Asefdjah in the production of GOLD, stage design: Petra Straß, 2020. ©Luise Rist

explorers of new female talent. We particularly support art made by women.
IRINA: That's wonderful.
17: It's really lovely to see how the second and third generations of immigrants are really finding their place in German society. Not merely in the world of work, but also in the art world.
18: I don't want to read what's written here.
1: When you think that in the countries of origin there's sometimes hardly any school education at all. Art teaching doesn't even get a look in!

The artist builds up contact with one of her objects. Without speaking, she enacts an art performance. The performances form one element in her planned production.

Migrational Female Artist = Migrational Art?

Schauspiel. Sie arbeitete als freischaffende Schauspielerin an verschiedenen Berliner Bühnen und ab 2019 auch für das »boat people projekt«. Sie spielte die Rolle der Irina Baryalei in der Produktion GOLD. Seit 2020 spielt sie mitunter am jungen Deutschen Theater in Berlin und ist ebenfalls im Bereich Film und Fernsehen tätig, 2021 war sie in dem Film NICO von Eline Gehring zu sehen.

Die Frage, ob eine migrantische Künstlerin migrantische Kunst machen muss – beziehungsweise ob ihre Kunst per se migrantisch ist, weil sie selbst migrantisch ist – kann ich

Javeh Asefdjah is an actor. After studying acting at the Felix Mendelssohn Bartholdy University for Music and Theatre, she worked as a self-employed actor on various Berlin stages and with the "boat people projekt" from 2019 – she played the part of Irina Baryalei in the production of GOLD. Since 2020, she has acted for the Young German Theater [Junges Deutsches Theater] in Berlin and has also worked in film and television. She can be seen in the film NICO *from 2021, directed by Eline Gehring.*

Responding to the question of whether a migrant artist has to make migrant art – or, put differently, whether her art is mi-

direkt mit einem NEIN beantworten.

Dies würde bedeuten, dass die Künstlerin beziehungsweise der Künstler lediglich auf ihre bzw. auf seine Herkunft reduziert wird. Eine solche Reduktion würde dieser Person absprechen, dass sie sich dafür entscheidet z. B. ein Bild zu malen, das ›lediglich‹ ästhetisch etwas zum Ausdruck bringen will. Es würde bedeuten, dass eine Künstlerin mit migrantischem Hintergrund auch nur Blumen mit migrantischem Hintergrund malen darf, um es humorvoll zugespitzt zu formulieren, sprich exotische Blumen auf deutschem Boden. In meiner Familie gibt es viele Künstler:innen, die ganz unterschiedliche Wege gegangen sind, um ihren Kunstwerken Ausdruck zu verleihen. Natürlich kann eine Künstlerin sich dafür entscheiden, ihre Herkunft und eigene Geschichte einfließen zu lassen, die Künstlerin kann sich dafür entscheiden, mit ihrer Kunst sozialkritisch und politisch zu sein, aber *müssen* tut sie das natürlich nicht. Als Schauspielerin spiele ich die bildende Künstlerin Irina Baryalei in dem von Luise Rist geschriebenen Theaterstück GOLD, und gleich zu Beginn des Stücks stelle ich die Fragen in den Raum bzw. lässt mich die Autorin fragen: »Wer legt wem etwas in den Mund? Was ist unaussprechlich für mich, was spricht durch mich? ... Wer ist das Ich, das da vor Ihnen steht?« Ich bin eine Schauspielerin, die sich dieser starken Frauenfigur mit all ihren Unsicherheiten zur Verfügung stellt. Zwar habe ich persönlich ebenfalls einen Migrationshintergrund, jedoch nicht den gleichen Hintergrund der Figur Baryalei. Mir sind also einige ihrer Gedankengänge sehr vertraut, wir teilen ähnliche Erfahrungswerte bezüglich

grational per se, because she's a migrant herself – I would utter a loud and direct NO!

To claim that this is the case would be reductionist, limiting the artist to their geographical origin. Such a reduction rejects the notion the person in question chooses to paint a picture that, for example, just wants to express something aesthetically. It would also mean that an artist who has autobiographical experience of migration is only allowed to paint flowers that also have a migrational background. In less serious terms, she would only be allowed to paint exotic flowers on German soil. In my family, many artists have taken entirely different paths, in order to lend a sense of expression to their art works.

Of course, a female artist can choose to incorporate her origins and history in her work; the same artist can opt to be socially critical and political through her art, but obviously she doesn't *have* to be. In my work as an actor, I play the visual artist Irina Baryalei in the play GOLD written by Luise Rist, and right at the beginning of the play I enter such questions into the theatrical space. Or, put differently, the play's author has me ask the following questions: "Who puts words into whose mouth? What is unsayable for me, and what speaks through me? ... Who is this me or this I, standing there in front of you?" As an actor, I put myself at the disposal of this strong female figure with all her insecurities. While having been a migrant is also part of my biography, my background is nonetheless quite different from the character of Baryalei. Yes, some of her thought processes are very familiar to me, and we share similar

der Reduktion auf Nationalität und Kultur; aber es reicht nicht, sich auf diesen Gemeinsamkeiten auszuruhen. Somit liegt es in meiner Verantwortung, mich tief mit ihrer speziellen Geschichte, mit ihrem Werdegang auseinanderzusetzen, im Prinzip wie bei jeder anderen Figur auch – nur, dass es bei Figuren mit einem Migrationshintergrund oft mehr, viel mehr zu entdecken gibt.

IRINA: Ich erinnere mich daran, wie erstaunt meine Eltern waren, dass es in den Häusern in Westberlin noch keine Heizungen gab. Dass ganz Berlin im Winter nach Kohle roch, dass die Häuser aussahen wie im neunzehnten Jahrhundert. Ich fühlte mich wie in die Zeit von Dostojewski zurückversetzt. – Und diese Kunstszene, die ganzen … renommierten Galerien …, die wir heute in Berlin haben, die gab es damals in Westberlin überhaupt nicht.
2: Sie wollen sagen, dass wir in Berlin keine Kultur hatten?
IRINA: Kunst. Nicht Kultur. Es gab keine richtige Kunstszene. Damals. Sind Sie Berliner?
3: Warum?
IRINA: Weil Sie »wir« gesagt haben.
4: Apropos Dostojewski. Als Künstlerin sehen Sie manches vielleicht aus einer etwas privilegierten Perspektive – nicht alle Flüchtlinge haben einen Zugang zu Literatur und Kunst. Wenn ich an die Tausenden von Menschen denke, die alle in Europa unterkommen wollen, da haben ja die wenigsten eine Chance bei uns auf dem Arbeitsmarkt. Da sehe ich schwarz.
IRINA: Ich habe selbst viele Jahre Probleme gehabt auf dem Arbeitsmarkt. – Auch ohne geflüchtet zu sein –

experiences of other people reducing us in terms of nationality and culture. But it is not enough to focus complacently on what we have in common. It is my responsibility to delve more deeply into her particular story, into her career, as I would, in principle, with any other character. It's just that there is often so much more to discover about characters who see themselves, or who others see, as migrants.

IRINA: I remember how amazed my parents were that there was no central heating in West Berlin houses and flats [at that stage]. That the whole of Berlin smelled of coal in winter, that the houses looked like they had built in the 19th century. I felt like I'd been transported back to Dostoyevsky's time. – And this art scene, all these … celebrated galleries … that we have in Berlin today didn't exist at all in West Berlin back then.
2: Are you trying to say that we people in Berlin had no culture?
IRINA: Art. Not culture. There was no real art scene. Back then. Are you a Berliner yourself?
3: Why?
IRINA: Because you said "we".
4: Speaking of Dostoyevsky. As an artist, you perhaps see things from a somewhat privileged perspective – not all refugees have access to literature and art. When I think of the thousands of people who all want to find their place in Europe, very few of them have a chance in our labour market. Then things start to look bleak.
IRINA: I had problems on the labour market for years myself – Without having had to flee myself –

Die Objektivierung von Migration

5: Na, da sehen Sie. Selbst für gebildete Flüchtlinge ist es nun mal nicht einfach.
6: Es heißt »Geflüchtete«.
5: Sage ich doch.
IRINA: Es geht um Erfahrung. Welche Erfahrung haben Sie gemacht, die in die Kunst einfließt? Das ist doch interessanter als die Frage, wo ich ursprünglich herkomme.
7: Bravo.
8: Sind Sie gläubig?
IRINA: Sie meinen, weil ich aus – weil meine Eltern –
9: Die Frage hat nichts mit Ihrer Herkunft zu tun. Ich sehe nur so viele religiöse Bezüge in Ihrem Werk.
IRINA: Ach so. Ja, Verzeihung, das hatte ich gerade falsch verstanden.
10: Das ist vielleicht eine zu persönliche Frage. Verzeihung.
IRINA: Nein, nein, das ist ... ich war nur gerade noch woanders, ich stand auf der Leitung. Ich glaube an eine Verbindung zu – ... zum Beispiel arbeite ich stark mit Fügungen, mit – diese Kunstwerke entstehen gewissermaßen auch im Dialog mit –
11: Herr Kurator! Wenn ich eine Frage an den Herrn Kurator stellen darf, der ja auch hier im Publikum sitzt: Hat diese Galerie einen Schwerpunkt für fernöstliche Kunst?
IRINA: Wie kommen Sie jetzt darauf?
12: Ich bin selbst auch in der Kunst tätig. Seit einiger Zeit habe ich keine Chance mehr, mit meinem deutschen Namen einen Job zu bekommen. Ich bekomme auch keine Förderung mehr. »Zu weiß, zu deutsch.« Was soll ich machen?
13: Was hat das mit Frau Baryalei zu tun?
14: Ich bin Schauspielerin. Ich bin arbeitslos. Weiße Frauen werden gerade nicht ge-

5: You see! It's not even straightforward for educated refugees.
6: We don't say 'refugees', we say 'people who have fled'.
5: That's what I said.
IRINA: What's at stake is experiences. Which experiences have you had that impact your art? That's much more interesting than the question of where I come from originally.
7: Hear, hear!
8: Do you believe in God?
IRINA: You mean, because I come from – because my parents ...
9: The question has nothing to do with your origins. I just see so many religious references in your work.
IRINA: Ah, oh, yes. My apologies, I misunderstood you just then.
10: That question was maybe too personal. Excuse me.
IRINA: No, no, that is ... I was just somewhere else in my thoughts, and so I was lost for words. I believe in a connection too ... I work intensively with acts of providence, for example – these art works also take shape, to a certain extent, in dialogue with –
11: The Right Honourable Curator! If I may ask a question to the Right Honourable Curator who is currently here in the audience: does this gallery focus on Far East art?
IRINA: How do you get that idea?
12: I also work in the art world. For some time now, I've no longer had a chance of getting a job with my German name. 'Too white, too German.' What am I meant to do?
13: And what's that got to do with Mrs Baryalei?

Abb. 4: Luise Rist, 2018. ©Alen Ljubic

Image 4: Luise Rist, 2018. ©Alen Ljubic

braucht, wurde mir gerade erst wieder letzte Woche von meiner Agentur gesagt. Das hat jetzt nichts mit Ihnen zu tun. Aber –

Die Künstlerin tritt in Kontakt zu einem ihrer Objekte. Kunst-Performance.

Die Verantwortung der Schreibenden

Luise Rist, Theater- und Romanautorin, Dramaturgin, Mitbegründerin von »boat people projekt« und regieführend bei vielen Projekten mit jungen Menschen verschiedenster Herkünfte und Muttersprachen.

Als Schreibende habe ich, bevor ich die erste Seite beginne, bereits mehrere Prozesse

14: I'm an actor. And I'm unemployed. Only last week my agency said to me that white women aren't needed at present. That doesn't have anything to do with you directly. But –

The artist makes contact with one of her objects. Art performance.

The Responsibilities of Writers

Luise Rist, Playwright and novelist, dramaturge, co-founder of the "boat people projekt", directing many projects with young people of different origins and mother tongues.

As a writer, I have already gone through several processes of approaching topics and people. But the moment I formulate the first sentence, I am alone. I also let people speak and talk about experiences that I haven't had myself. Every author can write about experiences they haven't had because the essence of art is that it is free and has to be free; literature, theatre, music and painting come into being because people move between worlds, and because those describing always occupy a somewhat marginal position. This is one perspective. But juxtapose that viewpoint with the artists who, because of worldwide structural racism against people of colour, do not have their say, are not heard, read or

der Annäherung an Themen und Menschen durchlaufen. Im Moment, in dem ich den ersten Satz formuliere, bin ich aber allein. Ich lasse auch Menschen sprechen, in denen von Erfahrungen die Rede ist, die ich nicht gemacht habe. Jede Autorin, jeder Autor kann über Erfahrungen schreiben, die sie oder er nicht gemacht hat, denn es liegt im Wesen der Kunst, dass sie frei ist und frei sein muss; Literatur, Theater, Musik und Malerei entstehen, weil sich Menschen zwischen Welten bewegen und als Beschreibende immer eine etwas randständige Position einnehmen. Einerseits. Andererseits gibt es Kunstschaffende, die aufgrund des weltweiten strukturellen Rassismus gegen People of Colour nicht zu Wort kommen, nicht gehört, gelesen oder gesehen werden, die aufgrund ihrer Hautfarbe oder Zugehörigkeit zu einer bestimmten Ethnie nicht die Möglichkeit zu einer künstlerischen Entfaltung bekommen haben. Ihnen den Vortritt zu geben, wenn es um die Einladung zu einem Podium oder zu einer Lesung geht, ist in diesem Kontext richtig. Als weiße, deutsche Autorin trage ich beim Schreiben zum Beispiel eines antirassistischen Theaterstücks eine besondere Verantwortung den Menschen gegenüber, die von Rassismus betroffen sind.

Über bestimmte Themen nicht zu schreiben, wäre hingegen nicht etwa übertrieben, sondern meines Erachtens falsch. Auch, weil es zu einfach wäre.

Es wäre ein fatales Signal, meiner Ansicht nach, wenn weiße Europäer:innen nicht über Themen wie Rassismus schreiben würden, aus Angst, in den Augen der »Anderen«, der Betroffenen, etwas falsch zu machen. Es ist nicht das Thema der »Anderen«, im Gegenteil ist es das Thema in erster Linie derer, die pri-

seen, and whose skin colour or membership in a certain ethnic group means that they have not been given opportunities to develop artistically. In this context, it is right to prioritise these people when it comes to inviting participation in a podium or in a reading. As a white, German author, when writing an antiracist play, for example, I have a special responsibility towards people who are affected by racism.

But *not* to write about certain topics would be, in my opinion, not merely an exaggerated step, but simply wrong. Wrong also because it would be too simple.

It would be a fatal signal, in my view, if white Europeans chose not to write about issues like racism for fear of doing something wrong in the eyes of the 'others', the people affected. This issue does not belong exclusively to these 'others', on the contrary, it is the issue primarily of those who are privileged, and who can make a difference for those affected by learning and communicating the knowledge of how to avoid discriminatory language and actions.

Regarding the question which role objects play in my engagement with flight and migration on the stage, I remember that the prop in my own first theatre production as a young student was my mother's small suitcase, the only piece of luggage she carried with her when she fled the GDR. A prop from one of the first plays we staged at "boat people projekt" with young people who had fled to Germany was a green and gold tea set, laid on a table that several participants carried onto the stage and which stood on a floor cloth representing the Mediterranean. A

vilegiert sind und die etwas für die Betroffenen verändern können, indem sie lernen und weitergeben, wie diskriminierende Sprache und Handlungen vermieden werden können.

Bei der Frage, welche Rolle Objekte bei der Beschäftigung mit Flucht und Migration auf der Bühne spielen, erinnere ich mich, dass das Requisit in meiner ersten eigenen Theaterinszenierung als junge Studentin das Köfferchen meiner Mutter war, das sie bei ihrer Flucht aus der DDR als einziges Gepäckstück bei sich trug. Ein Requisit aus einem der ersten Stücke, die wir bei »boat people projekt« mit nach Deutschland geflüchteten Jugendlichen in Szene setzten, war ein Set aus grün-goldenen Teetassen, die auf einen Tisch gedeckt wurden, den mehrere Teilnehmende auf die Bühne trugen und auf ein Bodentuch stellten, welches für das Mittelmeer stand. Ein Tee für die Toten wurde ausgeschenkt, dazu sang jemand ein Lied aus Eritrea.

IRINA: Ich habe für diese Performance, um die ich vom »Herrn Kurator«, der offenbar gar nicht anwesend ist, gebeten wurde, eine Schauspielerin gesucht. Für die Rolle der Papusza. Es war schwer, jemanden zu finden. Wissen Sie, warum? Es gibt kaum People of Color mit einer Schauspielausbildung. Heute ja, ja ja, jetzt sucht man die Stellen verzweifelt zu besetzen. Noch vor wenigen Jahren konnte man als Schauspielerin mit ›sichtbarem Migrationshintergrund‹ in überhaupt gar keinem Theaterstück spielen. Es sei denn, eine Reinigungskraft war zu besetzen oder eine Person mit kriminellem Hintergrund.
(*Zu Papusza*:) »Sie sind nicht gerade das Gretchen. Sie sind Fatma.«
15: Wo haben Sie eigentlich studiert? Oder ist das auch keine statthafte Frage!?

tea for the dead was served, accompanied by someone singing a song from Eritrea.

IRINA: For this performance, which the Right Honourable Curator, who's obviously not even here this evening, requested I put on, I looked for a female actor. For the part of Papusza. It was difficult to find someone. Do you know why? There are hardly any People of Colour who have been to acting school. Today, yes, oh yes, people search desperately to fill such positions. Up until just a few years ago, an actor 'visibly of migrant origin' could not act in any play at all. Unless the part being cast was a maid, or a person with a criminal background.
(*Too Papusza:*) "Right now, you're not playing Gretchen. You're Fatma."
15: Where did you actually train? Or is that question inadmissible too?
IRINA: Paris. *Académie des Beaux-Arts*.
16: Paris, mm-hmm!
IRINA: When I came back to Germany with my degree all signed and sealed, it was hard, bloody hard to find places to exhibit. To be honest, I didn't exhibit at all for many years. I was considered a kind of ... exotic delivery woman, I only received inquiries regarding art that had to do with my origins. The people who sat – and still sit – in the crucial positions were white, and they backed white men more than any other group.
17: Boring! Boring!!
18: A degree is not decisive in the art world.
IRINA: That's right.
1: Art [that's exhibited] has to blow its own trumpet.

Die Objektivierung von Migration

IRINA: Paris. Académie des Beaux-Arts.

16: Paris, aha!

IRINA: Als ich mit dem Diplom in der Tasche zurück nach Deutschland kam, war es schwer, verdammt schwer, eine Ausstellung zu bekommen. Ich habe, ehrlich gesagt, viele Jahre gar nicht ausgestellt. Ich galt ja als eine Art ... Exoten-Lieferantin, man fragte mich nur, wenn es um Kunst ging, die mit meiner Herkunft zu tun hatte. An den entscheidenden Stellen saßen – und sitzen – Menschen, die weiß waren und hauptsächlich weiße Männer gefördert haben.

17: Langweilig! Langweilig!!

18: Das Diplom ist ja in der Kunst nicht entscheidend.

IRINA: Richtig.

1: Die Kunst muss sich selbst behaupten.

2: Und in Frankreich? Warum haben Sie da nicht ausgestellt? Wenn man da multikultureller eingestellt war?

IRINA: Sehen Sie, Sie bringen mich wieder mit Multikulti in Verbindung.

3: An Ihrer Ausstellung sieht man einen Bezug zu orientalischen Themen. Das ist interessant, Ihre Perspektive ist bereichernd. Warum wehren Sie sich gegen Multikulti? Der Begriff ist etwas veraltet, gut, aber man muss ja nicht alles auf die Goldwaage legen.

4: Doch, muss man. Man muss die Worte abwägen.

5: Sie sprechen alle zu leise.

6: Wie bitte?

7: Ich finde es nicht richtig, was uns alles in die Schuhe geschoben wird. Es gibt auch Deutsche, die es nicht schaffen, in der Kunstwelt hochzukommen. Es ist immer schwer, hochzukommen.

8: Sie *ist* Deutsche.

IRINA: Danke. Wobei das auch total egal ist.

2: And what about France? Why didn't you exhibit there, if attitudes there were more multicultural?

IRINA: Do you see what you're doing? – You're associating me again with multiculti attitudes.

3: Viewers see connections to oriental themes in your exhibition. This is interesting, and your perspective is an enriching one. Why do you bridle against the attribute 'multiculti'? OK, the concept is slightly out of date, but does every single word have to be scrutinised before it is used?

4: Yes it does! Words must be examined carefully and compared to each other.

5: You're speaking too quietly.

6: Come again?

7: I don't find it right that we get blamed for everything. There are also Germans who don't manage to climb the ladder in the art world. It's always difficult to climb the ladder.

8: She *is* German.

IRINA: Thank you, even though that's totally irrelevant.

9: There's no question that not everything went right in the past, but that doesn't mean it's right to organise a witchhunt now against white men in leading positions.

10: That's reverse racism. People are discriminated against today because they are – yes, because they are white. That's racism, too.

11: Subsidies for art are being repurposed into a veiled subsidy of immigration.

IRINA: Racism against white people doesn't exist!

The Objectivisation of Migration

9: Sicherlich ist nicht alles richtig gelaufen in der Vergangenheit, aber man kann ja jetzt auch keine Hexenjagd auf weiße Männer in leitenden Positionen veranstalten.
10: Das ist umgekehrter Rassismus. Heutzutage wird man diskriminiert, weil man – ja, weil man weiß ist. Das ist auch Rassismus.
11: Kunstförderung wurde zur verkappten Förderung von Einwanderung umfunktioniert.
IRINA: Es gibt keinen Rassismus gegen Weiße!

Migrantische Figuren ohne spezifische Objekte

Waseem Alsharqi, in Deutschland lebender, syrischer Schriftsteller und Theatermacher. Er arbeitete als Dramaturg in verschiedenen Theaterprojekten in Syrien, der Türkei, Schweden und Deutschland. Seit 2020 ist er Masterstudent am Institut für Theaterwissenschaft der Freien Universität Berlin.

Ich persönlich glaube nicht, dass Erinnerungen bei irgendeinem Prozess des Theatermachens ausgeschlossen werden können. Erinnerungen sind eine Schlüsselkomponente in jedem kreativen Prozess, unabhängig vom eigenen Hintergrund. Ich denke jedoch, dass Erinnerungen in der Kunst von Menschen mit Migrationserfahrung eine größere Rolle spielen könnten, weil die Erinnerung meist das Einzige ist, das ihnen von ihrer ursprünglichen Heimat geblieben ist, und weil sie dort nicht mehr leben und höchstwahrscheinlich auch nicht mehr zurückkehren können.

Wie ein Dramaturg die schweren und harten Erfahrungen und Geschichten der Migration auf der Bühne darstellen kann, hängt

Migrant Characters Without Particular Objects

Waseem Alsharqi, a Syrian writer and theatre-maker based in Germany, has worked as a dramaturge in many theatre projects in Syria, Turkey, Sweden and Germany. He is currently studying for a master's degree at the Institute of Theatre Studies, part of the Freie Universität of Berlin.

I don't think that memory can be excluded from any theatre-making process. Memories are a key component in any creative process, regardless of the creators' backgrounds. However, I think memories may play a bigger role in art made by people who experienced migration, because memory is usually the only thing they still have from their original homes. And because they are not living in those homes anymore and most probably cannot go back again.

How dramaturges could represent the burdensome and difficult experiences and stories of migration depends on the context of each project. In a context where the dramaturge themself has personal experience with migration-related topics, they can help present that experience in a better way. For example, they can make sure that the narrative doesn't descend unwittingly into clichés or stereotypes.

I don't think we need some specific object to represent a migrant character on stage. Moreover, I think it's an unhealthy assumption to think that migrant characters on stage ought to be represented using particular objects. Individuals with

Abb. 5: Leen Hashisho in HEROINE, 2021. ©Sören Vilks   Image 5: Leen Hashisho in HEROINE, 2021. ©Sören Vilks

vom Kontext des jeweiligen Projekts ab. In einem Kontext, in dem der Dramaturg persönliche Erfahrungen mit migrationsbezogenen Themen hat, kann er helfen, diese Erfahrungen besser darzustellen. Er kann zum Beispiel dafür sorgen, dass die Erzählung nicht in Klischees oder Stereotypen abgleitet.

Ich glaube nicht, dass wir ein bestimmtes Objekt brauchen, um eine migrantische Figur auf der Bühne darzustellen. Außerdem halte ich es für eine ungesunde Annahme, dass Migrantencharaktere auf der Bühne mit bestimmten Objekten dargestellt werden sollten. Menschen mit Migrationsgeschichte haben Millionen von verschiedenen Geschichten und Leben vor ihrer Migration. Die Annahme, dass sie alle mit bestimmten Objekten oder Werkzeugen dargestellt werden können, öffnet der Stereotypisierung Tür und Tor.

migration stories have been through millions of different stories and lives before their migration. Assuming that all these lives and stories can be represented merely by using certain objects or tools just provides an opening to forms of theatre that stereotype the characters depicted in any given production.

VOICE: *Kaj san?*
IRINA: I'm sorry, but I really have to go. Is that the technician? Hello? To conclude this experiment, I want to tell you something personal. When coming here to you today, to this celebrated gallery, I got to know somebody, Danyar. I got into his taxi at the station …
VOICE: *Ada xira apr-o sundal, so mariben isi*
Pre manusenoär masa 'zdran, citron

STIMME: *Kaj san?*
IRINA: Ich muss leider abbrechen. Technik? Hallo? Ich möchte Ihnen zum Abschluss dieses Experiments etwas Persönliches erzählen. Auf dem Weg zu Ihnen, auf dem Weg zu dieser renommierten Galerie, habe ich jemanden kennengelernt. Danyar. Ich stieg am Bahnhof in sein Taxi ...
STIMME: *Ada xira apr-o sundal, so mariben isi*
Pre manusenoär masa 'zdran, citron
Aj 'gi rovel ratvale jasvenca
Nikom na zänel, kaj sie lesqre manusa
DANYAR: Salaam
IRINA: Guten Tag.
DANYAR: Wo kommst du her, Schwester?
IRINA: Berlin Paris Moskau, such dir was aus. Oder wie wäre es mit ... Paderborn ... »Bruder«?
DANYAR: Paderborn ist gut. Dein Deutsch auch.
IRINA: Weißt du nicht, dass das rassistisch ist?
DANYAR: Come on, Schwester. Wir sind Ausländer.
IRINA: Ich bin Deutsche.

Das Taxi gleitet durch die unbekannte Stadt, die Straßen wirken melancholisch durch die Musik, die im Wagen läuft.

DANYAR: Stört?
IRINA: Was? Die Musik? Nein. – Er dreht den Song ein bisschen lauter.
STIMME: *Nek del tumeque mro Devel baxt bari*
And-o ves kalo.
IRINA: Wir parken direkt vor der Galerie, der Fahrer öffnet die Fenster auf beiden Seiten, zieht seinen Mundnaseschutz herunter.
DANYAR: Danyar. – Mein Name bedeutet »der Weise«.

Aj 'gi rovel ratvale jasvenca
Nikom na zänel, kay sie lesqre manusa
DANYAR: Salaam.
DANYAR: Salaam.
IRINA: Guten Tag.
IRINA: Good afternoon.
DANYAR: Where do you come from, sister?
IRINA: Berlin? Paris? Moscow? – the choice is yours. Or what about ... Paderborn,[2] 'brother'?
DANYAR: Paderborn is good too. As is your German.
IRINA: Don't you know that's racist?
DANYAR: Come on sister! We're foreigners.
IRINA: I'm German.

The taxi glides through the unknown city, the music makes the streets seem melancholic, and the vehicle is moving.

DANYAR: Does that annoy you?
IRINA: What? The music? No. He turns the song up a bit.
VOICE: *Nek del tumeque mro Devel baxt bari*
And-o ves kalo.
IRINA: We park directly in front of the gallery. The driver opens the windows on both sides, and pulls down his face mask.
DANYAR: 'Danyar': My name means 'the wise.'
IRINA: OK, Danyar. This is an absolute no-waiting zone. He just grins.
DANYAR: Are you also one of those people who always does everything supercorrect, just to avoid looking like the stupid foreigner who can't follow the rules?
IRINA: I don't know what I'm meant to say. I can't be bothered with this type of

Die Objektivierung von Migration

IRINA: Ok – Wir stehen im absoluten Halteverbot, Danyar. Er grinst nur.

DANYAR: Bist du auch eine von denen, die immer alles super richtig machen, nur um nicht als dummer Ausländer da zu stehen, der sich nicht an Regeln halten kann? –

IRINA: Ich weiß nicht, was ich sagen soll. Ich habe keine Lust auf diese Art Gespräch, auf diesen Schulterschluss. Wir sind keine Opfer von niemandem, und ich kenne die Regeln im Straßenverkehr, weil ich genauso deutsch oder nichtdeutsch bin wie die, die da vom Ordnungsamt kommen und Strafzettel verteilen.

STIMME: *Ax, tu miri cerhenôrri! Angäl dives tu san bari.*

IRINA: Deine Musik ist super pathetisch, sage ich schließlich.

DANYAR: Gib zu, dass du das schön findest.

IRINA: Klar finde ich das schön. – Ich weiß nicht, worauf er hinaus will.

DANYAR: Mashallah ... Wir sind eben ein bisschen anders. Schönen Goldschmuck trägst du.

IRINA: Danke. Ich habe das extra angezogen, für meine Ausstellung, soll Glück bringen. – Anders als wer? Ich bin, ehrlich gesagt, geradezu deutscher als deutsch, wenn man mit Klischees kommen will, und ich bin in der Heimat meiner Eltern nie, hörst du, niemals, gewesen! Das ist eine lange, sehr politische Geschichte. Und, ach egal.

DANYAR: Deutsche Frauen finden diese Musik übrigens nicht schön. Sie finden sie nur kitschig.

IRINA: Die Musik ist kitschig.

DANYAR: Du magst sie trotzdem.

IRINA: Es gibt gar keine deutschen Frauen. Es gibt einfach Frauen.

DANYAR: Wie heißt du?

IRINA: Fatma.

conversation and this type of cosying-up. We aren't anybody's victims, and I know about traffic regulations, because I'm just as German – or not German – as the people who come from the Public Order Office and hand out fines.

VOICE: *Ax, tu miri cerhenôrri! Angäl dives tu san bari.*

IRINA: What I finally do say: Your music is totally melodramatic.

DANYAR: You find it beautiful, admit it.

IRINA: Of course I find it beautiful. I don't know what he's trying to get at.

DANYAR: *Mashallah* ... Me and you are just a little different. I like the gold jewellery you're wearing.

IRINA: Thanks, I put it on extra for my exhibition, because it's meant to be lucky. – Different than who? To be honest, I'm even more German than normal Germans, if we want to start bandying cliches about, and I've never been to my parents' home country – never, did you hear that! Which is one long and very political story. And – ah, forget it.

DANYAR: As it happens, German women don't find this music beautiful. They just find it kitschy.

IRINA: It *is* kitschy.

DANYAR: But you like it anyway.

IRINA: 'German women' don't even exist. There are just women.

DANYAR: What's your name?

IRINA: Fatma.

DANYAR: Really? You're just saying that to please me.

IRINA: Fatma pleases you?

DANYAR: Yes, of course.

IRINA: A really common name.

The Objectivisation of Migration

Abb. 6: Nora Amin, 2019. ©Jacob Stage

Image 6: Nora Amin, 2019. ©Jacob Stage

DANYAR: Echt? Das sagst du, um mir zu gefallen.
IRINA: Fatma gefällt dir?
DANYAR: Ja, klar.
IRINA: Heißen halt viele so.
DANYAR: Das gefällt mir gerade. –
Ist dir die Heimat deiner Eltern egal?
IRINA: Er macht den Motor aus, stellt die Musik ab. In meinem Kopf klingt die Melodie weiter.

Nein, natürlich nicht. Im Gegenteil. Aber das ist nichts, worüber ich mit Menschen, die ich keine fünf Minuten kenne, reden möchte.

Haare sind keine zum Opfer gemachten Migrant:innen; Haare sind ihr eigenes Heimatland

Nora Amin, ägyptische Autorin, Performerin, Tänzerin, Kuratorin, lebt aktuell in Berlin.

Erinnerungen sind für mich ein Raum für Wurzeln und Transformationen. Sie tragen

DANYAR: That's exactly what I like about it. –
Don't you care at all about the country your parents call home?
IRINA: He turns off the motor and the music. The melody rings on in my head.

Of course I care. But it's not something I want to talk about with people who I've only known for 5 minutes.

HAIR Is Not a Victimised Migrant; Hair Is Its Own Homeland

Nora Amin is an Egyptian author, performer, dancer and curator who currently lives in Berlin.

I see memories as a space of roots and transformations. They carry the beginnings and childhood, and transform with time so that the old memories themselves migrate within our minds and imagination and become different than their initial colour. On stage, I migrate to my own self-created memories and homeland, one I have selected myself. The stage is not a space for reminiscing but a platform to create new, evolving memories and identities. An arena in which wounds can be healed.

He hands me a business card with the number for his taxi company on it.

die Anfänge, die Kindheit, und verändern sich mit der Zeit, sodass die alten Erinnerungen selbst in unseren Gedanken und Vorstellung migrieren und eine andere als ihre ursprüngliche Farbe bekommen. Auf der Bühne migriere ich in meine eigenen, selbst geschaffenen Erinnerungen und in eine Heimat, die ich selbst gewählt habe. Die Bühne ist kein Ort für Rückbezüge, sondern eine Plattform, um neue, sich entwickelnde Erinnerungen und Identitäten zu schaffen. Eine Arena, um die Wunden zu heilen.

Er gibt mir eine Visitenkarte, auf der die Nummer seines Taxiunternehmens steht.

DANYAR: In meiner Heimat war ich Chemiker.
IRINA: Echt. Wow. Und jetzt …

DANYAR: In the country I come from, I was a chemist.
IRINA: Really? Wow. And now …?
DANYAR: Phone me when you need to pick me up in the taxi again.
IRINA: He refuses to take my money …
DANYAR: So I'll see you later on,
IRINA: … is all he says. As I open the door to the boot of his car, to get my suitcase, Danyar calls out:
DANYAR: What are you doing? Are you trying to steal my umbrella?
IRINA: Where is my suitcase?
DANYAR: You didn't have a suitcase with you, sister.
IRINA: That's when I realise that I've left my suitcase lying in the train. Totally on edge about this stupid exhibition, I forgot it completely. Shit. Danyar, what am I meant

Abb. 7: Yara Eid in der Produktion ›Herzlich Willkommen‹, 2020. ©Oktavia Ostermann

Image 7: Yara Eid in the production 'Herzlich Wilkommen' [A Warm Welcome], 2020. ©Oktavia Ostermann

The Objectivisation of Migration

DANYAR: Ruf an, wenn ich dich wieder abholen soll.
IRINA: Mein Geld nimmt er nicht an.
DANYAR: Wir sehen uns ja später ...
IRINA: ... sagt er nur. Als ich die Heckklappe seines Wagens öffne, um meinen Koffer herauszuholen, ruft Danyar:
DANYAR: Was machst du? Willst du meinen Regenschirm klauen?
IRINA: Wo ist mein Koffer?
DANYAR: Du hattest keinen Koffer, Schwester.
IRINA: In dem Moment realisiere ich, dass mein Koffer im Zug liegt. Vor lauter Aufregung wegen dieser blöden Austellung ... hab ich ihn komplett vergessen. Scheiße. Danyar, was soll ich machen, in dem Koffer war mein Kleid für heute Abend. Für meine Vernisage.

Erinnerungen auf der Bühne

Yara Eid, Tänzerin, Choreographin und Tanzpädagogin, geboren in Syrien, studierte Tanz in Damaskus und hat ihren Master in Tanzpädagogik an der Folkwang Universität der Künste in Essen abgeschlossen.

Die Kunst kann die Gesellschaft, die Menschen und die Natur spiegeln.

Aus meiner Erfahrung als Künstlerin kann ich auf der Bühne mit abstrakten Elementen arbeiten und ihnen die Bedeutung geben, die ich will und brauche.

Manchmal habe ich mich auf meine Erinnerungen – auf die Bilder, Musik und Geschichten in meinem Kopf – verlassen. Ich habe dieser Erinnerung vertraut und sie als Elemente für meine Ideen auf der Bühne benutzt.

to do, my dress for the vernisage this evening was in my suitcase.

Memories on Stage

Yara Eid, dancer, choreographer and dance teacher, born in Syria, studied dance in Damascus and completed her master's degree in dance education at the Folkwang University of the Arts in Essen, Germany.

Art can mirror society, the people in it and nature.

I draw on my experiences as an artist to work with abstract elements on stage and to give them the significance I want and require.

In this process, I rely partially on my memories – the images, music and stories in my head. I trusted this memory and used it as elements for my ideas on stage.

In my last dance production, *Herzlich Willkommen* [A Warm Welcome], staged by the "TheaterwerkstattW [Theatre Workshop] in Hannover, the focus was on memory and on the conflict between the past and the present. How strong a part does memory play in our present? Do we need these memories to 'climb aboard' into our present?

Do they help us with our future steps and plans?

She follows Danyar, who looks at the objects with a precise gaze.

Bei meinem letzten Tanzstück ›Herzlich Willkommen‹ mit der Theaterwerkstatt Hannover lag der Schwerpunkt auf der Erinnerung, auf dem Konflikt zwischen Vergangenheit und Gegenwart. Wie stark ist die Rolle der Erinnerung in unserer Gegenwart? Brauchen wir diese Erinnerungen um in unsere Gegenwart einzusteigen?

Helfen sie uns bei unseren zukünftigen Schritten und Plänen?

IRINA: Gemeinsam gehen wir in den Ausstellungsraum. –

Sie folgt Danyar, der sich die Objekte ganz genau ansieht.

Wird jemand sagen, dass das esoterisch ist, was ich mache? Esoterisch ist mein Reizwort. Esoterisch und exotisch ist das Werk einer von ... woher auch immer geflüchteten Künstlerin. Ich bin das nicht. Die ... geflüchtete Künstlerin.

Ich habe oft den Zwang, in Deutschland sagen zu müssen, dass ich aus einer Hochkultur stamme. Als ich klein war, dachten die Mitschüler, dass ich nicht mit Messer und Gabel essen kann. Das hatten ihnen ihre Eltern wohl über mich erzählt.

DANYAR: Es schmeckt besser, wenn man mit der Hand isst.

IRINA: Bist du verrückt? Sei still. Das nährt doch jedes Klischee.

DANYAR: Ich habe übrigens erst gedacht, dass du aus Afghanistan kommst.

IRINA: Aha.

DANYAR: Hast du etwas gegen Afghanen?

IRINA: Natürlich nicht. Ich weiß, dass viele Iraner auf Afghanen herabblicken.

Is someone going to say that the art I do is esoteric. Esoteric is a word that really triggers me. The work of a female artist who has fled from ... wherever ... is esoteric and exotic. That is not me. The ... female artist who has fled. In Germany, I often feel compelled to say that I come from a country with a high level of culture. When I was little, my classmates thought I couldn't eat with a knife and fork. I guess that's what their parents had told them about me.

DANYAR: It tastes better when you eat with your hand.

IRINA: Are you crazy? Just be quiet. That would just provide ammunition for every possible stereotype.

DANYAR: At first, I actually thought you were from Afghanistan.

IRINA: Uh-huh.

DANYAR: Do you have something against Afghans?

IRINA: Of course I don't. I know that lots of Iranians look down on Afghans.

DANYAR: But you still don't want to be an Afghan.

IRINA: I don't want to be Iranian either.

DANYAR: But it's better to type Isfahan than Kabul – is that how it works?

IRINA: What are you getting at?

DANYAR: You want to see people, but you only ever see what they've achieved culturally. That's very traditional middle-class.

Enter PAPUSZA.

IRINA: Danyar, I'm scared of this exhibition.

DANYAR: What makes me scared are the people who have spent longer periods

DANYAR: Du willst trotzdem keine Afghanin sein.
IRINA: Ich will auch keine Iranerin sein.
DANYAR: Aber besser man tippt auf Isfahan als auf Kabul. Richtig?
IRINA: Was willst du?
DANYAR: Du willst Menschen sehen, aber du siehst auch immer nur ihre kulturelle Leistung. Das ist auch sehr bürgerlich.

Papusza tritt auf.

IRINA: Danyar, ich habe Angst vor dieser Ausstellung.
DANYAR: Was mir Sorge macht, das sind die Leute, die mehrere Jahre im Ausland verbracht haben, und deswegen ihrer Meinung nach einen Antirassismusausweis mit sich herumtragen.
IRINA: »Ich bin farbenblind. Ich sehe quasi gar keine Hautfarben, ich sehe nur den Menschen. All Lives Matter!«
DANYAR: Das sind bürgerliche Leute, die sich weigern, etwas gegen die zu unternehmen, vor denen man dann richtig Angst haben muss –
   Mach dir keine Gedanken. Du hast deine Arbeit schon gemacht. Den Rest macht deine Kali-Göttin, ganz allein. Sie braucht uns nicht.
IRINA: Doch, sie braucht uns. –
Danyar, ich habe Angst, dass der Grund dafür, dass ich jahrelang nicht eingeladen worden bin, gar nichts mit meinem ausländischen Namen zu tun hatte. Sondern allein mit mir. Und dass der Grund, dass ich jetzt eingeladen wurde, wiederum nur mit meinem Namen zu tun hat.

abroad, and therefore think they're entitled to carry an 'antiracism pass' around with them.
IRINA: "I am colour blind. I don't see skin colours, as it were, I only see people. All Lives Matter!"
DANYAR: These are just middle-class people who refuse to act against those groups and individuals we really do need to be scared of …
   Don't worry yourself. You've got your job finished. Your Kali goddess will do the rest, all by herself. She doesn't need us.
IRINA: Oh, yes, she does.
Danyar, I'm worried that the reason I received no invitations to exhibit, for years, had nothing to do with my foreign name. But was rather just to do with me. And that the reason I've finally been invited to exhibit now has exclusively to do with my name.

Notes

1 Which carry the crude name in German of *Ausländerbehörden* (Immigration Office): authorities for foreigners

2 Paderborn, a small city of 150,000 citizens in central Germany, functions in many conversations in Germany as a humorous byword for provinciality.

Autor:innen / Authors

Joachim Baur ist Professor für Empirische Kulturwissenschaft an der TU Dortmund. Mit Katrin Pieper betreibt er die Ausstellungsagentur »Die Exponauten. Ausstellungen et cetera« in Berlin. Als Verbundpartner im Projekt »Zur Materialität von Flucht und Migration« kuratierte er die Webplattform und die begleitende Ausstellung MOVING THINGS.

Maliheh Bayat Tork ist wissenschaftliche Mitarbeiterin und Doktorandin am Institut für Ethnologie der Georg-August-Universität Göttingen. Als ehemalige humanitäre Helferin und Sozialarbeiterin in Flüchtlingslagern in Deutschland, Iran und Afghanistan untersucht sie in ihrer Forschung die Nachhaltigkeit der Auswirkungen und Ansätze humanitärer Infrastrukturen, die vom Globalen Norden zur Bewältigung der Migrationswellen aus dem Globalen Süden geschaffen wurden.

Seit 2009 arbeitet boat people projekt als Freies Theater in Göttingen. Sie produzieren in verschiedenen Konstellationen Schauspielproduktionen, Jugendstücke, Theatergames und experimentieren mit anderen Formaten wie Hörspaziergänge oder site specific theatre. Neben diesen Produktionen ist ihnen kulturelle Vermittlung ein besonderes Anliegen. Alle Projekt sind geprägt von Perspektivenvielfalt und Mehrsprachigkeit, auch das Publikum und die Teilnehmenden setzen sich aus diversen Gesellschaftsschichten und Hintergründen zusammen.

Joachim Baur is Professor of Empirical Cultural Studies at the Technical University of Dortmund. Together with Katrin Pieper, he runs the exhibition agency "Die Exponauten. Ausstellungen et cetera" in Berlin. As collaborator on the project "On the Materiality of (Forced) Migration", he curated the web-platform and accompanying exhibition MOVING THINGS.

Maliheh Bayat Tork is a research assistant and PhD candidate at the Institute of Social and Cultural Anthropology, University of Göttingen. As a former humanitarian and social worker in refugee camps in Germany, Iran and Afghanistan, her research investigates sustainability of impacts and approaches of humanitarian infrastructures established by Global North to deal with the waves of migration from Global South.

The boat people projekt has been operating as an independent theatre in Göttingen since 2009. They produce dramas, youth plays, theatre games with different line-ups and experiment with other formats such as 'audio walks' or site-specific theatre. In addition to these productions, cultural mediation is a major concern of theirs. All of their projects are shaped by diversity of perspectives and multilingualism, even the audience and participants come from a range of diverse social classes and backgrounds.

Peter J. Bräunlein ist Ethnologe und Religionswissenschaftler, der u.a. zu indigenen Kosmologien Südostasiens und materieller Religion in Museen forscht. Im Projekt »Zur Materialität von Flucht und Migration« war er als Senior Researcher tätig.

Elza Czarnowski ist Anthropologin und Museologin mit dem Schwerpunkt Critical Museology und Kommunikation. Sie ist seit 2018 im Rahmen des BMBF-Projekts »Zur Materialität von Flucht und Migration« tätig und hat anschließend als Teil des kuratorischen Teams die Ausstellung MOVING THINGS entwickelt.

Antonie Fuhse ist Ethnologin und war wissenschaftliche Koordinatorin im Projekt »Zur Materialität von Flucht und Migration« am Institut für Ethnologie der Georg-August-Universität Göttingen. In ihrer 2019 abgeschlossenen Promotion erforschte sie Mobilitätserfahrungen, die Aushandlung multipler Zugehörigkeiten und gesellschaftlicher Erwartungen.

Miriam Kuhnke hat einen Bachelor in Ethnologie und Allgemeiner Sprachwissenschaft der Universität Göttingen. Zurzeit ist sie Teil des ›Erasmus Mundus Master in Global Studies‹-Programms in Wien und Leipzig. Sie hat Feldforschung in Singapur durchgeführt und als studentische Hilfskraft im Projekt »Zur Materialität von Flucht und Migration« mitgearbeitet.

Andrea Lauser ist Professorin für Ethnologie an der Georg-August-Universität Göttingen. Themen in Forschung und Lehre sind (un-

Peter J. Bräunlein is a social anthropologist and religious scholar whose research interests include indigenous cosmologies of Southeast Asia and material religion in museums. He was a senior researcher in the project "On the Materiality of (Forced) Migration".

Elza Czarnowski is an anthropologist and museologist with focus on Critical Museology and Communication. Since 2018, she has contributed to the BMBF-Project "On The Materiality of (Forced) Migration" and subsequently, as part of the curatorial team, developed the MOVING THINGS exhibition.

Antonie Fuhse is a social anthropologist and was the scientific coordinator in the project "On the Materiality of (Forced) Migration" at the Institute of Social and Cultural Anthropology, University of Göttingen. In her PhD, completed in 2019, she researched experiences of mobility, the negotiation of multiple affiliations and social expectations.

Miriam Kuhnke has received a Bachelor of Arts in Social and Cultural Anthropology and General Linguistics at the University of Göttingen. Currently she is part of the 'Erasmus Mundus Master in Global Studies' programme in Vienna and Leipzig. She has conducted fieldwork in Singapore and worked as a student assistant in the project "On the Materiality of (Forced) Migration".

Andrea Lauser is Professor of Social and Cultural Anthropology at the University of Göttingen. Topics in research and teaching

ter anderem): Migration/Mobilität, Dynamik von Religion in Südostasien; »material turn« und Zukunftsvisionen. Sie war Sprecherin des Projektes »Zur Materialität von Flucht und Migration«.

Romm Lewkowiczs Forschung stützt sich auf politische Theorie, kritische Rechtsstudien und Ethnographie, um die Geschichte und Erfahrung von Migrationskontrolle und Bürokratie zu verstehen. Sein aktuelles Buchprojekt »Documenting the Undocumented« ist eine Ethnographie von Eurodac, einem Apparat zur biometrischen Kontrolle von unerwünschten Migranten in Europa. Er ist Research Fellow am Max-Planck-Institut für ethnologische Forschung.

Anoushirvan Masoudi ist Filmemacher, Medienkünstler und Medienwissenschaftler aus dem Iran. Seit 2020 promoviert er am Institut für Theater- und Medienwissenschaft an der Friedrich-Alexander-Universität Erlangen-Nürnberg. Fokus seiner Werke sind Themen wie Migration, Memory und Counter Memory, Ästhetik der Post-Medien und Iranisches Kino.

Veronika Reidinger, Studium der Soziologie und der Sozialen Arbeit, ist wissenschaftliche Mitarbeiterin am Ilse Arlt Institut für Soziale Inklusionsforschung und Lehraufträge am Department Soziale Arbeit, FH St. Pölten.

Özlem Savaş ist Kulturwissenschaftlerin mit Schwerpünkte Migrationskulturen und Medienpraktiken. Sie hat über kollektive und politische Gefühle der Migration, affektive digitale Medienpraktiken und diasporische

are (among others): migration/mobility, dynamics of religion in Southeast Asia; "material turn" and visions of the future. She was the spokesperson of the project "On the Materiality of (Forced) Migration".

Romm Lewkowicz' research draws on political theory, critical legal studies, and ethnography to understand the history and experience of migration control and bureaucracy. His current book project "Documenting the Undocumented" is an ethnography of Eurodac, an apparatus for the biometric regulation of Europe's unwanted migrants. He is a Research Fellow at the Max Planck Institute for Social Anthropology.

Anoushirvan Masoudi is a film maker, media artist and media researcher from Iran. Since 2020, he is working on his doctorate at the Institute for Theatre and Media Studies, Friedrich Alexander University of Erlangen–Nuremberg (FAU). The focus of his works are themes such as migration, memory and counter-memory, the aesthetics of post-media and Iranian cinema.

Veronika Reidinger studied Sociology and Social Work. She is a researcher at the Ilse Arlt Institute for Social Inclusion Research and has teaching positions at the Department of Social Work, St. Pölten University of Applied Sciences.

Özlem Savaş is a cultural studies scholar specializing in migration cultures and media practices. She has published on collective and political feelings of migra-

Ästhetik und Politik des Alltags veröffentlicht. Ihre aktuellen ethnografischen Forschungen befassen sich mit kollektiven Gefühlen, Reisegeschichten und Bezüglichkeit der Post-Gezi-Migration aus der Türkei.

Anne Unterwurzacher, Studium der Soziologie, ist wissenschaftliche Mitarbeiterin am Ilse Arlt Institut für Soziale Inklusionsforschung und Leiterin des Forschungsverbundes Migration im Rahmen des niederösterreichischen Forschungsnetzwerkes für Interdisziplinäre Regionalstudien.

Friedemann Yi-Neumann ist Ethnologe und war wissenschaftlicher Mitarbeiter im Projekt der Migrationsausstellung MOVING THINGS an der Universität Göttingen. Er arbeitet zu Flucht und Migration und Leben in Unterkünften für Geflüchtete aus einer Perspektive auf materiellen Kultur und home-making und mittels kuratorischer Ansätze.

tion, affective digital media practices, and diasporic aesthetics and politics of the everyday. Her current ethnographic research addresses collective feelings, travelling stories, and relationalities of post-Gezi migration from Turkey.

Anne Unterwurzacher studied Sociology. She is a researcher at the Ilse Arlt Institute for Social Inclusion Research and head of the Research Group "Migration", part of the Lower Austrian Reseach Network for Interdisciplinary Regional Studies.

Friedemann Yi-Neumann is a social anthropologist and was a research fellow of the migration exhibition project MOVING THINGS, University of Göttingen. He works on forced migration and life in asylum reception from a perspective on material culture and home-making and by curatorial approaches.

Impressum | Imprint

Herausgeber | Editors: Antonie Fuhse, Joachim Baur, Peter J. Bräunlein, Andrea Lauser, Friedemann Yi-Neumann

Lektorat | Academic Editorial Service: Henry Holland, Antonie Fuhse

Übersetzungen | Translations: Henry Holland, Antonie Fuhse, Andreas Hemming

Kataloggestaltung | Catalogue Design and Layout: Wallstein Verlag; vom Verlag gesetzt aus der Freight Text und der Montserrat

Umschlag | Cover Design: Susanne Gerhahrds, Düsseldorf, Foto: Hannah Bohr

Druck & Verarbeitung: Westermann Druck Zwickau GmbH

Diese Publikation begleitet die Ausstellung: This publication accompanies the exhibition:

»Moving Things«
Forum Wissen, Göttingen
Herbst 2022 | Autumn 2022
http://www.forum-wissen.de/

Kurator:innen | Curators: Joachim Baur, Elza Czarnowski, Miriam Trostorf, Friedemann Yi-Neumann, Antonie Fuhse

Gestaltung und Grafik | Exhibition and Graphic Design: Kooperative für Darstellungspolitik

Vermittlung | Art Education: Museum Friedland und Kunstverein Göttingen

Ausstellung und Katalog gingen aus dem BMBF-geförderten Projekt »Zur Materialität von Flucht und Migration« hervor, das von drei Verbundpartnern getragen wurde: dem Institut für Ethnologie der Georg-August-Universität Göttingen, dem Museum Friedland und dem Berliner Ausstellungsbüro »Die Exponauten. Ausstellungen et cetera«.

The exhibition and catalogue are the results of the BMBF-funded project "On the Materiality of (Forced) Migration", a joint undertaking by three partners: the Institute of Social and Cultural Anthropology at the University of Göttingen, the Museum Friedland, and the Berlin-based exhibition agency "Die Exponauten".

Unser Dank gilt den Förderern | We would like to thank our sponsors for their generous support:

  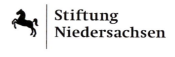

Weitere Informationen zum Projekt finden Sie unter | For more information about the project, visit:
https://materialitaet-migration.de/

Bibliographische Information der Deutschen Nationalbibliothek
Die Deutsche Nationalbibliothek verzeichnet diese Publikation in der Deutschen Nationalbibliografie; detaillierte bibliografische Daten sind im Internet über http://dnb.d-nb.de abrufbar.

Wallstein Verlag, Göttingen 2022
ISBN 978-3-8353-5190-5